Deb Hunt was born in England, where she worked as a librarian, teacher, event manager, PR executive, actress and journalist. She self-published her first book, *Dream Wheeler*, in 2013. Her second book, *Love in the Outback*, was published by Macmillan in 2014. She has worked with Shakespeare in the Park in London, *Australian House & Garden* magazine and with the Royal Flying Doctor Service. She has lived in France, Spain, Saudi Arabia, London, Broken Hill and a small village in Gloucestershire. Deb now lives in Sydney with her partner and their dog. She blogs at www.strawberriesinthedesert.com

Also by Deb Hunt

Love in the Outback

DEB HUNT

Australian farming families

Inspiring true stories of life on the land

MACMILLAN
Pan Macmillan Australia

First published 2015 in Macmillan by Pan Macmillan Australia Pty Ltd
1 Market Street, Sydney, New South Wales, Australia, 2000

Cataloguing-in-Publication entry is available
from the National Library of Australia
http://catalogue.nla.gov.au

Typeset in 12/16 pt Fairfield LH Light by Post Pre-press Group
Printed by McPherson's Printing Group
Map by Laurie Whiddon

The author and the publisher have made every effort to contact
copyright holders for material used in this book. Any person or organisation
that may have been overlooked should contact the publisher.

MIX
Paper from
responsible sources
FSC
www.fsc.org   FSC® C001695

For my family

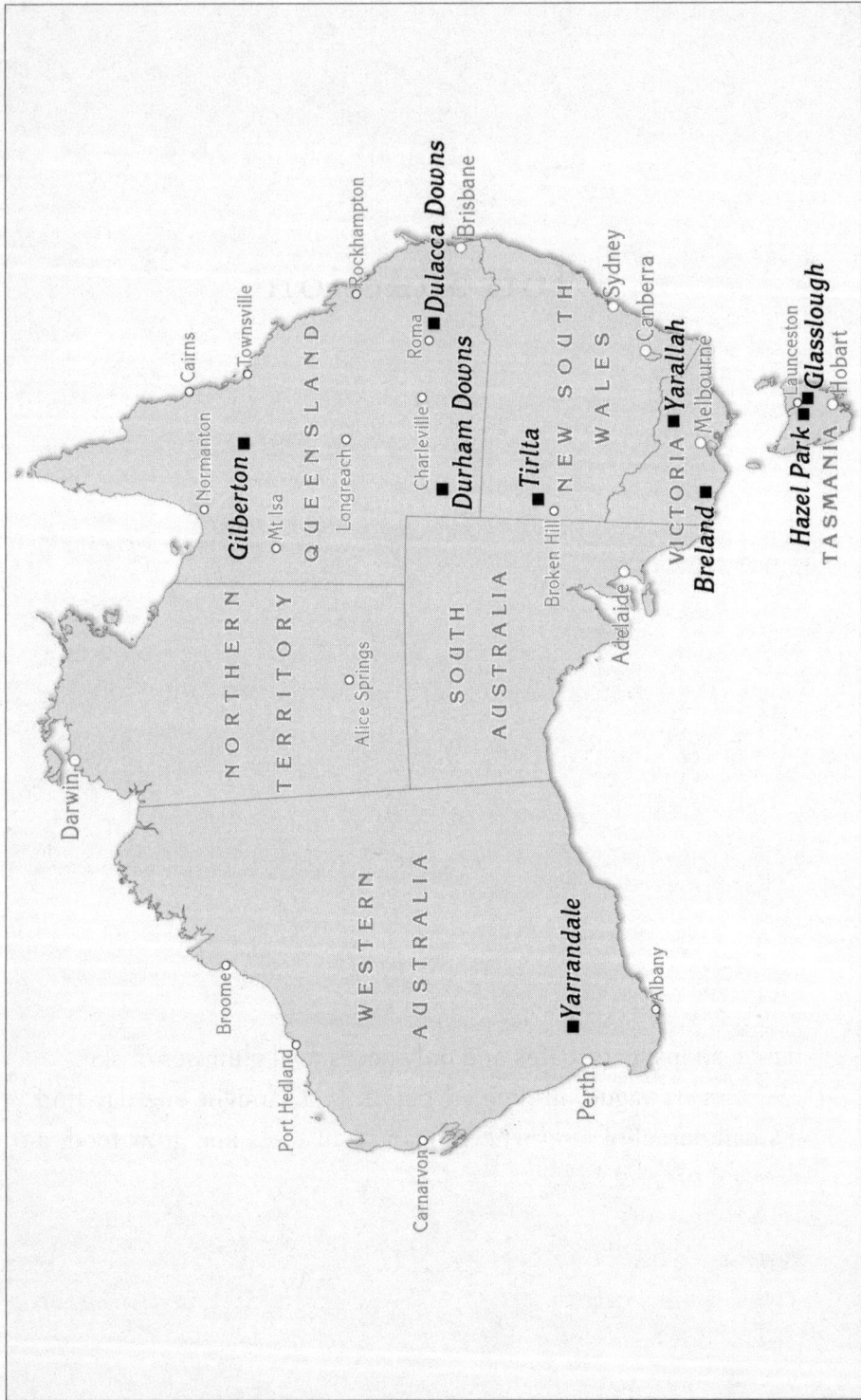

Introduction

I have a confession to make – I'm hopelessly sentimental when it comes to land. I long to stand in a paddock under a gum tree, inhale the scent of eucalyptus and listen to the wind rustling through leaves. I want to tramp across the bush, watch eagles fly overhead and feel the energy of the earth vibrating through my boots. It's a hopelessly romantic view, I know, a visceral desire, when the reality is that home for me is a densely populated suburb of Sydney's inner west.

It's a place of parks and trees – even odd glimpses of water – yet all I notice are tightly packed terraced houses crammed onto tiny plots, concrete patios, bitumen-covered pavement, traffic lights, shops, cars, cafes and only occasional glimpses of sky.

I carry a vague, ill-defined notion that I might one day find a small parcel of land where I can plant trees and grow food, a dream I suspect I might share with a not inconsiderable number of other people. It hasn't happened yet. I live in the city and dream of open space.

From the outset I was an unlikely candidate to write a book about farming families; I have no background in farming and I've never lived on the land. Leaving aside my lack of farming knowledge, I also have no children and I was born in England, not Australia. It's more normal for me to refer to a paddock as a field and to land as countryside. Lastly, I'm a lifelong vegetarian. And as if all that weren't enough, I'm hampered by my absurdly romantic view of farming.

Other than a few years living in Broken Hill, where I grew vegetables and kept chooks, I had little connection with Outback Australia when I embarked on this project. Writing a book about life on the land has challenged my stereotypical views and my ignorance about farmers, graziers and pastoralists (like the fact that there's a difference between them for a start).

I accepted the perceived view that 'bushies' were people whose pragmatic stoicism got them through the tough times they often faced, although what those tough times were was only vaguely articulated in my mind: something to do with fire, flood and drought. I've since added frost, hail, dust storms, pests, disease, debt, accidents and illness to that list.

Just as I was doing my best to try to bury my romanticism under a flurry of pragmatic, unemotional facts on farming, two parcels arrived from the Galls at Langawirra, friends who run a sheep station northeast of Broken Hill. Inside the first parcel were two small ziplock bags; one contained sandy earth and the other fine red dust. The second parcel contained a bouquet of everlasting daisies, neatly held together by a rubber ring normally used to castrate sheep.

Lynne Gall had collected the red earth from one of the paddocks at Langawirra, and she'd scooped up the dust from a truck stop on the outskirts of Broken Hill, where drivers stop to let the dust fall from their truck rims before they drive into town. Her grazier husband, John, picked the everlasting daisies from a paddock after a brief shower of rain.

The contents of those two parcels astonished me. The Galls had always struck me as a classic bush couple – hard-working, pragmatic people with not a hint of sentiment or romance about them.

'Hope you enjoy this little touch of the bush,' Lynne wrote. 'If you want, you can sprinkle it [the red earth] like fairy dust on your city garden.'

The unexpected gifts gave me a glimmer of hope that my sentimental attachment to land wasn't so misplaced after all. Those everlasting daisies bloomed for weeks in my small studio as I worked on this book.

It takes a special kind of person to live happily away from the facilities most of us take for granted, someone like Cath Marriott, who lost her husband to cancer shortly after his fortieth birthday and who was left to run their Victorian sheep farm and raise their four young children alone, or Lyn French, who ran away from her strict father in Queensland when she was fourteen and by fifteen was working full time as a roustabout.

Then there's Roma and Glenn Britnell, who were knocked back time and again when they tried to borrow enough money for a sheep property, so they proved the banks wrong by switching from sheep to cows and they built their own dairy by hand. Or what about Ian Jackson who was, on his own admission, the most feral kid ever to draw breath and the 'dumbest bastard' in school? He went on to meet the President of the United States of America. Michelle Reay arrived in Australia as an English backpacker fresh out of university and somehow ended up with four sons, living and working at Durham Downs, often referred to as the jewel in the crown of the Kidman cattle empire.

All of the interviewees I visited had fascinating stories to tell and they lived and worked in unique places, from the rugged high country of the goldfields of northern Queensland to the grassy slopes of Tasmania; from the dusty red earth bordering Mutawintji National Park to the vast productive paddocks of Western Australia's central wheatbelt.

Most of us cluster near the coastline in this vast continent of ours. We turn our faces to the ocean and our backs to the land, rarely seeing what goes on beyond the coastal fringe. The Outback can seem a grim, forbidding place, thousands of kilometres of seemingly empty space with no mobile phone coverage and the nearest major hospital many hours drive away. Perhaps that's why the Royal Flying Doctor Service is such a well-respected and much-loved organisation – we all know how frightening it would be to have someone you love stranded 'out there', desperately ill and waiting for help to arrive.

Having worked for the Royal Flying Doctor Service for several years as a writer I had some idea of the challenges people faced in remote areas. I learnt a lot more writing this book. There were times during the research and interview process when I questioned my sanity, especially when I found myself driving alone on a dirt track hundreds of kilometres inland with nothing but wild pigs and kangaroos for company. More often than not, though, I felt blessed to have been granted the opportunity to find out more about the lives of people who live and work on the land.

Food is fundamental to our health and we can't survive without it, but in our neatly packaged, over-processed, technology-driven lives we're in danger of forgetting where it comes from. Food doesn't come from supermarkets; it comes from the land, and unless we're prepared to grow our own, the food we eat in every café, every restaurant and every home has been planted, nurtured and harvested or bred, fed and raised by people who live and work on the land.

The harsh reality of farming in the twenty-first century, according to much of what I read and many of the families I interviewed, is that a lot of people who live and work on the land today make less per hectare, per cow, per sheep or per pretty much anything they plant or raise than their parents, or sometimes even their grandparents, did.

But this book isn't about profitability or margins. I know more about farming now than I did before (and believe me, that wouldn't have been hard) but if you're reading this to learn something about farming techniques you're going to be disappointed. This book is about the stories that lie behind farming; about the people whose lives are intrinsically linked to the welfare of the animals they raise and bound up in the health of the land they farm.

What interested me most during the interview process were the human aspects of farming – the stories of success and failure, of life and love, of hardship and celebration – and the passion and gritty determination that characterised every family I talked to, from those who have lived on the land for generations to those who are relatively new to it.

The eight families featured in this book welcomed me into their lives with unconditional warmth and hospitality (even after I'd confessed to being a vegetarian) and although we had never met before, and in most cases only spoken once briefly on the phone, each family invited me to stay with them. It was a humbling experience and I'm not sure I would have found a similar welcome from strangers in a city.

I've done my best to keep a lid on any overtly sentimental musings in the writing of this book (and I suspect I haven't always succeeded).

Any mistakes are mine. Any credit for a story well told belongs to the families I interviewed.

Deb Hunt

Philip and Adele Hughes

'Dulacca Downs'
90 kilometres east of Roma, Queensland

The unlikely pairing of Adele and Philip intrigued me. Adele was born in Melbourne in the early 1950s, the daughter of a well-connected family whose grandmother travelled as a companion to Sir Robert Menzies' wife, Dame Pattie. She was sent to finishing school and she trained as a home economics teacher: hardly the sort of background you'd expect of someone who would go on to marry an uneducated stockman who'd left school at fifteen after he failed grade ten. Philip Hughes, though, was no ordinary stockman.

The fifth generation of an originally Welsh farming family – hence the surname – Philip's thirst for knowledge and his hunger to succeed saw him read every book he could get his hands on after he left school. It came as no surprise to learn that Philip had risen from humble jackaroo to head stockman then manager at some of the most successful properties in the Stanbroke Pastoral Company.

The stories they told me around their kitchen table at Dulacca Downs, when the paddocks outside were white with frosted grass and drought still gripped that part of inland Queensland, had me weeping with laughter and sometimes just weeping. Tales of their early life together called to mind the stories of Henry Lawson or the bush ballads of Banjo Paterson, hinting at a bygone era peopled with drovers and swagmen, stockmen and cooks, jackaroos and itinerant gardeners – characters who craved the isolation of remote settings and who didn't 'fit' into the system. As someone who struggled to find his place in school, Philip knew how that felt. He struck me as a manager who would be slow to judge and quick to forgive.

The couple's eventual decision to set up in business for themselves at a stage in life when they were comfortably established towards the top of the corporate tree spoke volumes about their entrepreneurial spirit and their drive to succeed. Their undeniable grit and vision, coupled with a strong bond of love and a great sense of humour, have seen them through some of the darkest times imaginable. They're living proof that persistence pays.

Their paddock-to-plate beef production business – Rangeland Quality Meats – was only set up in 2011 but it's already been recognised for excellence within the industry. Their oldest son, Lachlan, joined the business a few years ago and moved to an adjoining property with his wife, Anna, and their son, William. Since I met them, Adele and Philip's second son, Alister, has also joined the family business.

They've gone through the usual ups and downs of any father/son working relationship, plus the mother-in-law/daughter-in-law dynamic, and they've come through it stronger, wiser and more connected than ever.

Lachlan and Anna's commitment to living sustainably with as light a footprint as they can has driven Adele and Philip even further in the direction of organic farming, improving the soil that supplies the grass that feeds the cattle that, week after week, are sent to market to keep RQM at the top of its game. It's a game where the pieces have to fit together like a 3D puzzle to ensure a consistent supply.

Theirs is a family bond of love and laughter that has been strengthened through hardship and isolation, through success and failure, and through decades of sheer bloody hard work.

Philosophical Farmer

I met Philip Hughes at Roma airport, 470 kilometres inland from Brisbane. It occurred to me as I disembarked from a plane full of miners – including several women – that I hadn't told Philip what I looked like, nor had I thought to ask him the same question. I needn't have worried; Philip's blue shirt, dusty boots and battered Akubra instantly identified him, no doubt as instantly as my idea of city clothes suitable for the bush must have identified me.

It was generous of Philip to drive ninety kilometres to pick me up, even more generous of him and his wife Adele to invite me to stay for two days of interviews, but if he was anxious at the prospect of a writer delving into their lives he didn't show it. He seemed open and friendly, a loose-limbed, gently spoken giant of a man who found the prospect of being interviewed faintly amusing. I, on the other hand, was a bag of nerves. This was the first of eight visits I was planning to make to farming families across Australia and my cupboard was bare when it came to knowledge about cattle; even preparing basic questions had been beyond me. I had an empty tape recorder and two days of interviews ahead of me.

We were driving east, towards Dulacca Downs, an 8000-acre property where Philip and Adele had lived for the past ten years. I guessed Philip was probably in his late fifties, and it struck me as unusual that he and Adele had only lived at

Dulacca for such a relatively short time. This was my first wrong assumption: that people living on the land don't move around much.

Gentle drops of rain hit the windscreen as we drove out of Roma, past sprawling camps of flimsy dongas and white cabins that had been thrown up to accommodate the thousands of largely single men and women working in the coal seam gas industry; it was the modern equivalent of a gold rush.

'Roma was one of the first places they found oil originally, way back around the turn of the last century,' Philip said, resting his huge forearms on the steering wheel as he scanned the ranks of white dongas that marched across the landscape.

I assumed he must have been worried about the impact of mining on the grazing country we were driving through. But his concerns went deeper than that.

'When you have a big bucket of money and you fly in, do your thing then leave again, that has an impact on families. I think that's when we lose sight of what community means. In functioning communities it's not about everybody getting on, it's about accepting that there are differences and that you're all in it together.'

I learnt on the drive east that Philip had lived and worked in enough diverse communities to know what he was talking about. As we left the dongas and, sadly, the rain behind, he gave me a quick rundown. At fifteen he'd started work as a jackaroo and by eighteen he was head stockman at Dixie Station, west of Cooktown. He'd worked his way onto larger stations near Normanton, Winton and Windorah, and by the age of twenty-five he was managing 28,000 hectares at Banchory Station near Clermont, for the Stanbroke Pastoral Company. Two years later he was sent to Augustus Downs with the same company then to Bulloo Downs, a vast property of more than a million hectares carrying 20,000 head of cattle. I had no idea where any of these places were, and I hadn't given much thought to the possibility of people 'managing' stations rather than owning them either.

It sounded like Philip had been a rising star in one of the most iconic organisations in the Australian cattle industry (even I'd heard of Stanbroke) then he and Adele had left it all behind and struck out on their own.

As Philip described some of the setbacks, disasters and challenges they had faced since going it alone, I couldn't help thinking they might be regretting their decision. They weren't young, it sounded like they had struggled to make ends meet and their business was far from booming. My head spun with unfamiliar names and places. Then Philip was talking about hybrid vigour, pH levels, ticks and backgrounding, feedlots, grain-fed accreditation and I was forced to confess my ignorance of all things related to cattle.

'Well, it's like this . . .'

I let the conversation unfold about the various merits of different breeds of cattle – *bosincas? bostaurus?* – as the kilometres rolled past, trying desperately to remember if I'd told Adele I was a vegetarian. Then Philip let out a short bark of laughter.

'You'd have to say we've been on a long and rocky road with a few unexpected detours up some pretty dry gullies.'

His philosophical turn of phrase captured my attention.

'With the gadgetry we've got now I can pull up here on the side of the road, hop onto my iPhone and see how much meat they sold today. I can tell you exactly what's in my bank account, just from the side of the road. But is that a good thing? We live at quite a frenetic pace now so it's pretty important to keep in touch with nature to slow down a bit.'

I'm with you on that one.

'But it's all good. If you can put a family business together where each member can spread their wings, find things they like doing and build off the back of one another, you'll be ahead of the game.'

The 'game' Philip and his family were in was beef cattle, a paddock-to-plate production business that to his obvious pleasure now involved the whole family, including Lachlan and Alister.

'It hasn't been easy and there's no pot of gold in it,' he said as we turned off the main road onto a dirt track. 'As a parent you sometimes think it's your children's responsibility to come back and work, and your right to tell them what to do.' He let out another explosive rumble of thunderous laughter as we pulled up outside a simple homestead, shaded by mature trees and surrounded by paddocks of what looked like dry, brittle grass.

'I've sure had to pull my head in on that one!'

He was still smiling when Adele came out to meet us.

'Welcome to our humpy,' he said.

*

The first thing that struck me about Adele was how immaculately well groomed she looked, from her shining white hair coiled into a neat bun on top of her head to the tips of her manicured fingernails; next was the difference in height between her and Philip (she barely reached his shoulder); and finally, how frequently she (and he) laughed.

From the little I knew about them they seemed an unlikely couple, and the more I found out the more unlikely their pairing seemed. Adele was born on the outskirts of Melbourne. Apart from having a grandmother (still alive at 103 at the time of writing) who travelled as a companion to Dame Pattie Menzies during Sir Robert's tenure as Australia's longest-serving prime minister, Adele was schooled at Presbyterian Ladies College in Melbourne and later trained as a home economics teacher.

'And did she tell you she went to finishing school?' Philip shouted from his office, across the hall from the kitchen. 'She likes to keep that last bit quiet.'

Adele laughed.

'It's true, I did. And I always said I would send any children I had to a co-ed school because when I came out of school I had a strange attitude to boys, it was "Wa-hey! Boys! Where are they?"'

She was still laughing as she reached into the oven for a batch of white chocolate chip and macadamia nut biscuits.

'Nice nails,' I said, with genuine admiration. My own were splintered and grubby by comparison.

Adele smiled. 'They're fake, my only indulgence.'

A farmer's wife with fake fingernails? *They're not farmers, they're graziers. Okay, a grazier's wife with fake fingernails, still unusual surely?*

And how to reconcile Adele's immaculate appearance with the cabinets in her cramped kitchen that would have been ripped out and refurbished by any style-conscious Melburnian long ago?

'We'll probably shift the whole house onto the ridge, there's no point refurbishing it.'

It felt like Adele had read my mind and found it amusing; her breezy self-confidence suggested she wasn't remotely concerned what I thought of her kitchen, and rightly so.

'Did Adele tell you I was chosen as the first ever school captain at Charleville School of the Air?' Philip asked, reaching for a cookie as he strode into the kitchen.

'And did Philip tell you he was the only kid *in* grade seven?' she countered.

They both roared with laughter and I began to relax. The next two days promised to be a lot of fun; Adele and Philip Hughes may have been the most unlikely couple on paper but they were a natural pair in person.

'I was at cross-purposes with the system,' Philip explained, as we sat at the dining room table over morning coffee. 'With animals, some of your best are the ones that are hardest to deal with. Same with humans. If you have a human with a brain who won't be dominated, you can't thrash him into shape and stick him in a box, you have to find another way to reach him.'

Philip had followed a long line of Welsh ancestors onto the land. Five generations back, HB Hughes set up on the Eyre Peninsula, one of four brothers who made the journey to Australia, and he made his fortune carting freight on paddle-steamers up the Darling, buying properties with the proceeds. Philip's side of the family was descended from the youngest Hughes brother, who came out last and ended up in Queensland. Philip's father, George Lucas (Bill) Hughes, and his grandfather, Henry Maddock Lucas Hughes, managed Nockatunga near Thargomindah for a total of fifty-seven years between them.

Finally, a name I recognised! Nockatunga was north of Tibooburra, a small town I once visited when I lived in Broken Hill. The long, largely dirt road north from Broken Hill to Tibooburra ended in a settlement of houses with not much more than a pub, a village hall and a small hospital. In 2014 Tibooburra was the hottest place in New South Wales.

Philip may not have had much formal education but he had a thirst for knowledge, and working on stock camps had given him all the time and opportunity he needed to slake that thirst. 'We only went to town maybe twice a year and there was limited radio, so that left plenty of time for reading. Everything I learnt, I learnt through books. It was a wonderful way to get knowledge.'

Philip grew up a bushman through and through. He learnt how to read the land and how to understand the rivers that criss-crossed the landscape like lines on the palm of his hand. Coloured pins stuck into a map of Queensland hanging on the wall behind him indicated the many stations where he and Adele had lived and worked. I soon realised that Philip could describe it all without any reference to the printed map; his mental map was far more detailed.

'That's my backyard,' he said.

I wondered how on earth an educated young woman from Melbourne – who once dreamt of opening a gourmet bread and cheese shop – had ended up marrying a 'bushie'?

*

It was the late 1970s and Adele was in her mid-twenties, still living at home with her parents. She'd spent five years as a home economics teacher at a Baptist girls grammar school and all her spare time, including weekends and holidays, with friends on properties. She loved the bush, she loved horse riding and she was itching for a more exciting life. The answer was to offer her services as a governess in a remote area.

The first reply to her ad came from a place called Darkwood, inland from Coffs Harbour, near Bellingen in New South Wales. The handwriting was immature, as if the letter had been dictated to a child, so Adele's strict father rang the local minister and discovered Darkwood was a hippy commune. He vetoed it.

The second letter was from a Lutheran school – out of the frying pan and into the fire – and the only other reply came from Terrick Terrick, an 80,000-hectare sheep station not far from Blackall in Queensland, almost 2000 kilometres north of Melbourne.

The station manager's wife, Mrs Ruth Harvey, wrote a long letter describing how all the young people on the station played tennis and went horse riding . . . 'and there are only nine children to teach', she added.

Adele didn't stop to consider the demands of teaching nine children in five different primary school grades (made all the more challenging given her inability at maths). She simply thanked her lucky stars that she wouldn't have to teach classes of thirty and she would be able to ride a horse. With the naïve enthusiasm of the young, she took the job.

Most inner-city parents would have been concerned about sending their child to a remote sheep station, even if that child was now an adult in her mid-twenties, and Adele's were no different. What worried them most was the threat of snakebite. Anti-venom at the time contained a high percentage of horse serum and Adele had developed an allergy to horse dander, so for Christmas that year her parents bought their daughter an unusual present: a twelve-gauge shotgun. Adele broke the shotgun into pieces and packed it into her suitcase, tucked under her neat little schoolteacher outfits.

Station manager Peter Harvey met Adele off the plane in Blackall, where the dry heat hit her forcibly.

'Man, that's hot!'

Peter smiled. Adele wasn't the first city girl to arrive and she wouldn't be the last. 'Is there anything you need while we're in town?'

'Maybe some chalk and pencils,' she said, suppressing a desire to ask for cigarettes. 'And I need to get to the police station.'

'Why?'

'I want to register my shotgun.'

Peter dutifully took the diminutive young governess to the local station and stood behind her as he addressed the sergeant over the top of her head.

'This is our new governess,' he said gravely then paused, before adding in a deadpan voice, 'she wants to get her shotgun registered.'

By the time Adele was introduced to the jackaroos, the stockmen, the ringers and the overseer at dinner at Terrick Terrick that night – all of them gorgeous young men – the story had spread. One young stockman shook her hand and introduced himself.

'G'day, Annie.'

Adele assumed he had misheard and politely corrected him. 'It's Adele.'

'Ah, I thought it was Annie, as in *Annie Get Your Gun*.'

Before long everyone in town knew not to mess with the new governess who owned a shotgun, not that Adele ever fired it. The only time it came out of her suitcase was at weekends, when she threw dried cowpats into the air so trigger-happy boys could blast them out of the sky. When Adele tried it she almost blew her shoulder off.

Admitting she could type, if only with two fingers, landed Adele the job as secretary of Blackall's polocrosse club, with responsibility for organising the annual carnival dinner. Philip was a keen player and arrived at the dinner late, after spending the night in the pub with his team. Adele was no stranger to the pub herself but she was still put out that Philip and his players hadn't bothered to attend her carefully organised dinner.

'You have to pay full price to get in,' she insisted.

No amount of sweet-talking would sway her, and Philip and his team were rewarded for their lateness with cold spaghetti and a band that had finished playing.

Their next meeting was at a polocrosse carnival at Thylungra, a sheep station owned by the famous Durack family, and this time they managed a dance. Adele showed her hand without meaning to the next day when she sauntered past a young jackaroo.

'Who was that tall dark handsome guy I danced with last night?'

'You could ask him yourself.'

To Adele's acute embarrassment, Philip was sitting quietly beside the jackaroo.

The carnivals and the dances kept coming until one night at the Sand Goanna Slide, at Yaraka, Adele watched Philip dance with someone else. His partner was a very pretty girl in a red dress ('much slimmer than me') and Adele couldn't help noticing he had his hand firmly planted on her bum.

She marched across, grabbed the youngest, greenest jackaroo who worked for Philip and slow danced with him in front of them.

It worked.

*

'Have you seen that mole on Philip's back?'

It was just after Christmas, late December 1979, and Adele was standing in the kitchen at Fort Constantine near Cloncurry, the station Philip's parents had managed since they'd left Nockatunga. Apart from the occasional letter, Adele and Philip hadn't had any contact with each other since the end of the polocrosse season, several months earlier. Adele had mentioned in one of her letters that she was thinking of going back to Melbourne to open a gourmet bread and cheese shop and Philip's response was to invite her to spend Christmas at his parents' place.

'My mother's not what you would call a touchy-feely person,' he said by way of preparation.

Originally from Sydney, Phyllis Hughes had worked as an Australian Inland Mission nurse at Dunbar Station, a cattle property on the southern side of the Mitchell River in the remote cyclone-prone area of Far North Queensland, way up in the middle of Cape York Peninsula.

A practical woman, Phyllis realised when she took the posting that it would be some time before she might see a dentist, so she had all her teeth extracted before she left. She also spent a useful couple of hours learning how to perform extractions on other people, training that came in handy when her future husband, Bill, arrived at the clinic to have one of his own teeth extracted.

Now there she was, a dour, upright woman, not given to laughter, standing at the ironing board in the kitchen at Fort Constantine, asking Adele what seemed like a leading question. Had she seen the mole on Philip's back?

Adele's Christian upbringing kicked in. What had Philip said about her? What was his mother asking? Her mind raced through the possibilities. You don't have to get naked to see a mole on somebody's back, what if he'd simply had his shirt off? She'd worked with him, she could have seen it.

'Yes,' she said, eventually.

'And what did you think?'

'My mother always told me that a black mole wasn't good and you should have it removed.'

'Well maybe you can talk to him because I've tried and his sister has tried and he won't listen.'

Adele persuaded Philip to get the mole checked, and a biopsy at Princess Alexandra Hospital in Brisbane revealed third-stage melanoma. Several days later the mole was successfully removed.

In between the two hospital appointments Adele took Philip down to Melbourne to meet her parents, a clear sign that the relationship was becoming serious.

'What are you planning to do next year?' Philip asked after their brief visit to Melbourne.

'I thought I'd open that cheese shop I told you about.'

'We won't see much of each other if you do that.'

'I could fly up for a few polocrosse carnivals.'

There was a pause, but if Adele was hoping for a romantic proposal she was disappointed. 'Maybe we should just get married.'

She thought about it for a while. 'Yeah, well I suppose so,' she conceded.

Adele's parents were quietly relieved that their 26-year-old daughter had finally met someone she wanted to settle down with, although the implication of living and working in a remote area didn't really sink in for any of them.

After a period of recuperation, Philip splashed out on a new car – a bright yellow Toyota SR5 with a double cab – and the pair set off to drive 1500 kilometres due west of Brisbane, heading for Tanbar Station near Lake Yamma Yamma, in the hot and desert-dry Channel Country. Tanbar Station was almost as far as you could go in southwest Queensland before hitting the border with South Australia. At twenty-five Philip was working as head stockman at Tanbar for the Stanbroke Pastoral Company and Adele had been appointed governess.

Philip made a rare diary entry in January 1980 that read:

Arrived back at Tanbar after a good holiday with a new car, 30 stitches and a fiancé [spelt finance].

Their wedding day, in July 1980, prompted another fulsome entry:

Got married.

*

Philip and Adele's first few years of married life read like an epic bush ballad. (Who am I kidding? Most of their married life sounded like a Banjo Paterson poem to me.) First they moved from Tanbar to Banchory Station, 1000 kilometres northeast, where Philip was appointed manager and Adele worked as cook. The job of preparing a cooked breakfast, lunch, dinner and two smokos a day for all the staff who worked on the 28,000-hectare station meant early starts and long hours, but Adele still found time to do the thing she loved, which was to jump on a horse and pitch in with mustering. There was nothing prissy about Adele Hughes, no matter what her background might have been; she slept in swags with the best of them.

From there they were sent to Augustus Downs, way up in the Gulf of Carpentaria, ninety kilometres inland from the chunk of coastline that looks like the gap left when a tooth has been pulled. It was frontier country, wild and different, with big open plains and flat grassland. Darwin was 1200 kilometres in one direction, Brisbane 1700 in the other and Adele was six months pregnant.

Augustus Downs was an ailing property, neglected and badly run, and a far cry from its heyday in 1946, when the valet to the visiting Duke of Gloucester – then Governor-General of Australia – took a photograph showing a two-storey well-maintained building with neatly clipped hedges, painted verandas and shades at the windows.

Adele arrived with Philip in January 1982, on a hot, steamy day at the height of the wet season, to find the station deserted. The stock camp at Augustus Downs closed at the start of the wet season in November and reopened in March, so apart from Jack – a reliable man who had been head stockman at Fort Constantine and who now had responsibility for checking bores at Augustus Downs – and his partner Bernice, they were on their own.

Goats roamed freely through the muddy garden and the station yard was full of chooks. Empty ice-cream containers had been dug into the dirt outside the back door for chooks to lay eggs in, and there was chicken shit everywhere.

Philip led the increasingly astonished Adele through the homestead and into the office, where he pointed at a two-way radio.

'That's how we communicate,' he said. 'Our call sign is 8XF.'

Augustus Downs held the distinction of being the first station to send a telegram to a Flying Doctor base via two-way radio on 19 June 1929. Philip's parents had a connection with the Reverend John Flynn, who started the service in 1928, that went back to the day he officiated at their wedding. Philip knew how much they would have to rely on the Flying Doctor in such a remote spot. He and his siblings had been evacuated many times from Nockatunga as children, including once when he'd developed a severe allergic reaction to yabbies and again when his four-year-old sister, Wendy, suffered a burst appendix.

Adele was due to give birth for the first time in less than three months, and she was cut off from her parents as well as her friends. She had no access to antenatal clinics, no local GP and no phone. Her only communication was with the Royal Flying Doctor Service and a handful of other people via two-way radio, on an open channel with others listening.

It was a completely new experience for her.

Drinking water came from rainwater tanks – and there was no shortage of that when over 600 millimetres (almost twenty-four inches) could fall in a season – but water for general domestic use was piped from the nearby Leichhardt River. For six months of the year the river was crystal clear. Come the wet season the Leichhardt churned into a muddy red soup.

Pipes feeding water to the house ran above ground, so not only was the water unbearably hot, it was also unusable until it had been settled overnight in troughs. Adele learnt to add Epsom salts to help settle the sludge before inserting a tube into the top layer of clear water and sucking to make it flow. If she made the mistake of swallowing she spent the rest of the day on the loo.

Philip meanwhile had discovered a cold room filled with grog. When staff returned from their annual break it was clear they were used to an open-slather policy. He cut the ration to a more moderate six-pack a night (per person).

Among the staff were several 'cowboy' gardeners, who tended the vegetable garden, milked the cow and looked after the chooks. From Adele's description they sounded like the tramps and drifters I've often seen (and often overlooked) sleeping rough or living on park benches in a city; out there they were a vital part of a functioning cattle station.

Philip's management style was firm but fair. He knew from experience how to bring out the best in people and his philosophy was simple: put someone in a role where he's struggling and he'll get upset and cranky, put him in a role where he's achieving and he'll grow.

To overcome the isolation of such a remote spot Philip insisted that everyone, including pilots and gardeners, attended regular polocrosse carnivals, often held hundreds of kilometres away. Nowadays it would be called team building; to Philip it was simply a way for his station 'family' to get together and have some fun.

The person Adele was most eager to meet was the station cook, Peter Cornelius. Peter had spent his holiday, as he did every year, at Quamby, a tiny town on the road to Cloncurry where he had an arrangement with the publican. Peter would place his cheque, representing all the money he'd earned that season, on the bar and the publican would cash the cheque then eke the money out, supplying food, grog and a place to stay until it was time for Peter to go back to work again a couple of months later. The critical component in the equation was grog.

The prospect of having a cook on a station with twenty-five staff was a huge relief for the heavily pregnant Adele and she was eagerly waiting to meet him when he arrived.

'I knew he wouldn't be wearing a chef's outfit but even so, his appearance was still a shock.'

Adele watched an elderly man with grey hair and a white beard stagger out of the car. Peter Cornelius was wearing dark boxer shorts, thongs and a t-shirt full of holes that said 'Where the hell is Quamby?' He had the worst varicose veins Adele had ever seen and he looked to be in his early eighties. When a three-legged dog jumped out of the car behind him Adele burst into tears.

'I cried a lot.'

Peter Cornelius may not have been what Adele was expecting in a cook but he taught her valuable lessons about large-scale catering. An ex-merchant navy seaman, originally from Bristol in the UK, he was an educated man and an extremely capable cook who always kept a supply of corned beef or cooked meat in the fridge so anyone could make themselves a sandwich. He also made the best fruitcake, which Adele quickly learned was far better to take mustering than biscuits that broke.

*

I have no direct experience of pregnancy and childbirth. My three sisters, who do, all lived within a twenty-minute drive of the nearest maternity ward, so I can only imagine how worried a first-time mum would be living a five-hour drive from the nearest hospital on roads that were impassable in the wet season. Adele's sheltered upbringing hadn't given her the first clue about what to expect from childbirth, she hadn't attended a single antenatal class and she had no phone to contact friends or family for advice; what's more, any question she asked on the two-way radio at Augustus Downs was heard by everyone listening on the open channel. So who could she ask for advice?

'The *Everywoman Gynaecological Guide for Life*. That book was my bible. I read it from cover to cover.'

We were chatting over yet another coffee in the kitchen at Dulacca Downs (with yet more homemade biscuits for sustenance) and laughing about Adele's panicky conviction that she had every ailment going. Philip's answer was to hide the book under a carpet and plead ignorance when she went looking for it.

Adele's concern seemed perfectly understandable to me. I once offered to babysit a friend's newborn and felt rising panic when the baby wouldn't feed. Was the milk too hot? I let it cool down then worried that I couldn't safely reheat it. Could I feed a baby cold milk? And why wouldn't it stop crying? The helplessness I felt that day, faced with such a simple task, was overwhelming.

Adele's nearest hospital was in Mount Isa, and the only way to get there in the wet season was a ninety-minute flight on the station plane. I thought about those times and distances as Adele recounted her story; it was the equivalent of flying from London to Paris for an antenatal check-up, or from Sydney to Melbourne. Residents living in any Australian city would be horrified at the thought. For women in Adele's position, there was no option.

Thankfully, Adele's fears were largely unfounded, although at a final check-up in Mount Isa she was diagnosed with oedema and admitted to hospital. Lachlan was born two weeks prematurely on 28 March 1982.

Not every woman in that small Mount Isa Hospital was so fortunate. After the birth Adele vividly remembered being taken to a ward with three beds; on one side of her was a woman who lost her baby at birth, on the other a woman whose newborn son died after a short time in a humidicrib.

'I selfishly thought, why me?'

Not wanting to distress the grieving mothers, Adele went out of the room to feed her newborn baby on the veranda, where she sat with local Aboriginal women and listened to the advice of various nurses and aides on how to deal with cracked nipples.

'Put wool fat on them.'

'That's too wet.'

'Dry them with a hairdryer.'

'Try cabbage leaves.'

The one person Adele longed to ask was her mum, who lived several thousand kilometres away in Melbourne. Philip redeemed himself after the incident with the *Everywoman* by arranging a surprise visit from Adele's mum, flying her up to the remote mining town so she could spend time with Adele after the birth.

Philip left them together and drove back to Augustus Downs, but he saw them again sooner than anyone expected. Within hours of arriving at Augustus he was on his way back to Mount Isa, this time as a patient.

'He was thrown from his horse yarding cattle and he broke his ankle when it got stuck in the stirrups,' Adele explained.

Still on a drip, she waited in casualty for him to arrive.

'When I heard the ambulance driver shout, "Come and give us a hand with this fella, he's hanging out the back of the ambulance!", I knew it had to be Philip.'

The break was a bad one, although Philip was more concerned about the expensive boot they had to cut off, and he was kept in hospital for several days.

'The nursing sister said she'd never seen a better way for a new dad to get to spend time with his wife and son.'

I had made a conscious decision before I arrived at Adele and Philip's house that I wouldn't probe for any dark secrets or tragic events to 'dramatise' their story. Having thrown away any hope of asking meaningful interview questions, I decided that whatever came out in the course of our two days together would form the basis of what I wrote. As it happened, Adele didn't need any prompting to recall events that touched their lives with tragedy.

For all his abilities as a cook, it was obvious Peter Cornelius wasn't a well man. One of his jobs was to turn on the generator at five o'clock each morning, a task that powered up everything else on the station, and Adele and Philip sensed the day would soon come when they would wake up without power. Twenty months after Lachlan's birth, in late November 1983, Peter Cornelius hit the grog hard at the station Christmas party.

The next day Philip went in search of their errant cook and found him still in bed, frothing at the mouth. In spite of his obvious sickness he insisted on going to Quamby as usual and he suffered another turn at the pub. He died in hospital a couple of days later.

Philip and Adele had known Peter Cornelius less than two years, but in that short time he had become part of their station family. He'd proved himself a reliable cook and a gentle, kindly man. Both Philip and Adele had grown fond of him. As a manager on a remote station Philip was often the only constant in some people's lives, and he accepted it was part of his job to look after people like Peter.

Like many old-timers, Peter Cornelius had kept his past to himself and they knew little about him. Adele went through his few belongings to see if he had any family. In the end it took help from Interpol to track down relatives in England.

The funeral was to be held in Cloncurry, a three-and-a-half-hour drive from Augustus Downs, on a day when Far North Queensland sweltered in temperatures that hit 120 degrees

Fahrenheit in the shade. The whole station attended – ringers, gardeners, cooks, jackaroos and stockmen – dressed in an assortment of ill-fitting jackets with ties too short, coats too small and trousers too long. They were there to show their respect and to farewell a friend.

The mourners gathered under a tree opposite the church, in what Adele remembered was the hottest part of Cloncurry, watching a big storm brew in the west while they waited for the undertaker to deliver the coffin. Adele was pregnant with her second child, Mary, and the stifling heat must have been hard to bear. The undertaker eventually drove up in a big car and parked on the other side of the road. To everyone's surprise he didn't bother to get out of the car. Adele watched in heat-stupefied amazement as he wound down the window and shouted across the road.

'Won't be any funeral today. Stupid bloody nurses laid him out wrong. Crossed his arms. If they'd laid him out right he'd have frozen right. Only way we can get the bugger in the box now 'ud be to break his arms. Stupid bloody nurses should have known better. You'll have to come back tomorrow.'

Adele's voice rose as she recounted the story, her anger at such disrespect for a man they considered part of their family only just below the surface.

'Can you imagine if any member of his own family had been there? It was horrible. That undertaker just wound his window up and drove off.'

The local minister came to the rescue with a memorial service and they held a brief wake in the pub before Philip organised for the rest of the staff to go home. He and Adele stayed for the burial, rescheduled for the following day.

'Got him thawed out, got him in the box now,' said the undertaker, which earned him a blistering tongue-lashing from Adele.

'How dare you be so disrespectful! This man was like part of our family and you will treat him with the respect he deserves!'

Her outrage was wasted on the indifferent undertaker. 'What are you doing now?' he asked when she started collecting cards from the few simple wreaths.

'They are to send to his family,' she said patiently. 'To show them that people cared for this man.'

'Don't know why you're bothering, those cards just normally rot and blow away.'

And with that, the undertaker climbed into his old Fiat from the 1930s, and drove off.

'A hateful man,' she said, flatly.

We'd been talking for several hours by now and I wondered if Adele wanted a break. She shook her head.

'I haven't told you about Mary yet, have I?' she said quietly.

'No, you haven't.'

'Well, a few weeks after Peter Cornelius died, I was back in Mount Isa Hospital.'

Adele must have gone over the events she went on to describe many times, wondering if anything could have been done differently, if she should have been sent to hospital earlier, if she should have asked for help sooner, if miscommunication with the Flying Doctor on the two-way radio might have led to the series of errors that culminated in her giving birth to her daughter on the fifth of January 1984, at just twenty-eight weeks. Wherever the truth lay, Adele had long since worked through her grief and she wasn't about to apportion blame.

'I wasn't meant to have a daughter,' she said. 'I was meant to have two sons.'

Mary was born in a critical condition and placed on life support at Mount Isa Hospital while Adele and Philip spoke to a visiting paediatrician from Brisbane. He gave them grim news. Their daughter was unlikely to survive beyond the critical first few weeks. Even if she did, she was likely to suffer considerable disability.

Before they'd had Lachlan, Adele and Philip had talked about how they might cope if they had a disabled child in a remote area. It was a difficult conversation and they'd resolved to wait and see, to seek medical opinion, and only then decide what to do. The hospital suggested they take twenty-four hours to make their decision.

What would I have done in Adele's place I wondered? I've often wished I'd had children, but I'm fortunate that I've never had to face such a heartbreaking choice. Having lived in Broken Hill and worked for the Royal Flying Doctor Service, I knew how limited medical facilities must have been in a remote place like Augustus Downs in the mid-1980s. How would they have coped with none of the support systems people took for granted in urban areas? They had no family within easy reach and no hope of seeing a speech pathologist, an audiologist or an ophthalmologist. There were no specialised health profes-sionals like music therapists or physiotherapists and they were a five-hour drive from the nearest major town, which itself had limited facilities. Their only medical care was courtesy of the Flying Doctor.

'We talked and talked. Halfway through the night we made a long-distance call to my parents and we decided to switch off the life support machine.'

Adele didn't sleep that night. She went and stood in the dark-ness under a white frangipani tree blossoming in the hospital courtyard and noticed a flower lying on the ground. She picked it up, went back in and handed it to a nurse.

'I couldn't speak, I was beyond words.' The nurse understood. Next time Adele went in to see Mary, the white frangipani flower was lying in the humidicrib with her.

'I thought back to those women in hospital with me when I had Lachlan. At the time I thought, why me? Why me between these two women? I didn't keep in contact with them and now I knew what they had gone through.'

An Anglican minister conducted the simple burial service and Philip laid his daughter in the ground himself, in a small white coffin, at Mount Isa cemetery.

The next mail truck to arrive at Augustus Downs delivered an unexpected gift – a toy lawnmower, ordered as a present for Lachlan by the lovely Peter Cornelius just before he died.

'So much is interlinked in the bush. Other people had lost babies and they got through it. They went on with life.'

Adele did the same. Six months later she fell pregnant again, an ectopic pregnancy this time, and eventually she gave birth to their second son, Alister, in January 1986.

New cooks came and went, including the voluptuous Catherine from Swan Hill, the argumentative Beryl, Brenda who lived on packet soup, Cathy who couldn't cook and had to be sent home after a week, Sally the pathologist, Mary from Melbourne who wore Adele Palmer shirts at the stove, Margie who couldn't make tea and Cindy who only lasted a month.

No one, though, could ever take the place of Peter Cornelius.

*

The family moved next from the steamy heat of Augustus Downs to the red sands and flat plains of Thargomindah in southern Queensland, close to the border with New South Wales. Turning the ailing Augustus Downs property around in less than four years had earned Philip a promotion to manager on the million-plus-hectare Bulloo Downs, a vast property that carried 20,000 head of cattle.

Their new home was in a desert climate, in low sandy country regularly washed by floods from the Bulloo River and marked by open woodland of mulga and gidgee. Under the surface lay the vast water supplies of the Great Artesian Basin.

One of Adele's toughest challenges in living so far from Melbourne, especially once she had children, was that she rarely saw her parents. She hated not being able to get in a car and pop over to see them. When her mother was diagnosed with amyloidosis – a rare disease that affected her kidneys – it was even harder.

For once, I knew how Adele must have felt. Both my parents were still alive when I immigrated to Australia and I remember the feelings of guilt and sorrow when my mother was taken into hospital in England: the long-haul flight back, the weeks sitting by her bedside knowing I would soon have to return to work and commitments on the other side of the world. The prospect was unbearable and I left only when she seemed to be rallying. Days later I had a call to say she had worsened. My sister's face at Heathrow airport when I landed the second time told me I hadn't made it back in time. It was a miserable period, repeated when my father died.

On the rare occasion when her parents managed to visit, Adele would tell her children to treat their grandmother carefully. 'You have to treat her like a little thing, be gentle with her,' she would say. The nickname stuck, and 'Little Thing' would often be found sitting on the floor playing teddy bear picnics with her grandsons, her clothes elegant and her nails immaculate, before excusing herself to undergo another bout of dialysis.

Alister lay in bed one night listening to his mum read him a story after one such visit.

'What will we do when Little Thing dies?' he asked.

'We might talk and we might cry a bit.'

'No,' said her youngest son. 'She wouldn't want us to cry.'

At least they had a phone at Bulloo Downs so they could stay in touch with each other. Barbara's kidney transplant helped and, in Adele's opinion, her mother's strong Christian faith sustained her, adding a few more years, but with the onset of cancer the inevitable end came far sooner than anyone would have liked. Adele left the children with her mother-in-law while she visited her mother in hospital in Melbourne, and she stayed as long as she could before she had to leave. She'd got as far as Mildura when her father rang with news of her mother's death.

In their second year at Bulloo Downs, in what Philip referred to as a 'blending of the waters', his father and uncles bought Banchory – the cattle station near Clermont that he and Adele had managed almost a decade earlier. Bill Hughes, senior pastoral inspector for the Stanbroke Pastoral Company, was now part-owner of Banchory.

There was a lot of discussion about whether Adele and Philip would leave the relative security of Stanbroke to join the new operation. Bulloo Downs was one of Stanbroke's least successful properties when Philip had been appointed and it was now fast becoming one of their most successful. The following year they would top the company league table for sales turnover and send off 10,000 bullocks for an average dressed weight of 340 kilograms. Quitting the corporate world wasn't easy, as Philip explained.

'Managing a big property was fascinating. You were pretty much lord and master; you didn't play that card, obviously, but you did control everything, from employing staff to dishing out the beer ration and deciding when the generator got turned off.'

The downside of life as a manager was what would happen later.

'If you lived in a corporate world, sooner or later you would retire and then it would be too late to set up on your own. You can sometimes spot old managers, they wither on the vine when they leave those big properties.'

In the end, education drove their decision to quit Bulloo Downs.

'Patience for other people's kids wasn't a problem' said Adele, 'I could be sweet and kind and understanding, but when it came to my own . . . I was just so *bad* at it.'

It was difficult to believe that this engaging, funny, light-hearted, gregarious woman – who had been a teacher and governess – would have found it hard to home-school her own children.

'Believe me, I was bad.'

And she was.

A few days after Easter, in April 1987, Adele was trying to teach Lachlan his colours. The schoolroom with its thick mud brick walls was hot and stuffy after the long break and five-year-old Lachlan was wriggling in his chair, frowning at the desk.

'Blue,' he said, in answer to his mum's question.

'No, Lachlan, that's green. Try again. What colour is this?'

Lachlan scuffed his toes against the floor.

'Stop fidgeting and tell me what colour it is.'

'Blue.'

'No.'

'Red.'

'No!'

Adele marched into the house, grabbed a handful of mini Easter eggs with brightly coloured wrappers and stormed back to the schoolroom. She slapped them down on the table in front of her terrified son.

'Right! Every one of those you can tell me the colour of, you get to eat!'

Lachlan was so distressed at this point he didn't even hear his mother's tempting offer and burst into tears and the lesson finished early. Having tried and failed to be a primary school teacher (I lasted less than a month on a teacher training course in London) I sympathised with Adele's lack of patience.

The prospect of Mistake Creek Primary School being just twenty-five kilometres from Banchory swung their decision; buying into the family business meant the boys could go to a 'local' school.

The arrangement worked while the boys were in primary but Adele had always known there would come a day when her sons would have to go to boarding school, which didn't make it any easier when that day arrived for Lachlan. She dutifully sewed on name tags, ironed clothes, packed a trunk and delivered her eleven-year-old son to his new school in Rockhampton, where she was horrified to see one father shake his son's hand at the school gates.

'I hugged Lachlan, drove around the corner and burst into tears. The sight of this forlorn little thing standing outside the school, waving goodbye, was awful. It felt like someone had pulled my arm off and thrown it away.'

Five hours later Adele arrived back at Banchory, where she unleashed her pent-up distress on Philip. 'It's all your fault,' she sobbed. 'If I hadn't married you I wouldn't have had to have my children torn away from me!' Philip sensibly waited until the storm had passed before trying to comfort his distraught wife.

On free weekends, Philip and Adele would grab a quick shower after finishing work on Friday then jump in the car and drive east. Several hours later they would roll out a swag, catch a few hours sleep and complete the drive to Rockhampton early the next morning.

'We'd watch the kids play cricket, take them out somewhere for the day, drive back home on Sunday and do it all again a few weeks later.'

My occasional early-morning stroll down to Birchgrove Oval to watch cricket constituted a big fat lazy lie-in by comparison.

*

It wasn't until the second day of my visit that Philip and Adele drove me through the paddocks at Dulacca Downs. To my inexperienced eye the grass looked pale, dry and brittle. I may not have known anything about cattle but having lived in Broken Hill I thought I knew something about drought.

'Hmm, looks like you're in the grip of drought,' I offered.

'That's not dry, it's frosted.'

'Ah.' That explained the bitterly cold feeling I'd had when I'd gotten up that morning then. With the naïve ignorance of a city dweller I had assumed Queensland would be warmer in winter than New South Wales, as if somehow the temperature I'd checked for Brisbane before I left might have been indicative of the entire state. As a result I'd been forced to don every item of thin, entirely inappropriate clothing I'd brought with me.

Philip pointed to a smattering of green shoots close to the ground.

'See that clover? That's a staple favourite for the cattle.'

My eyes looked beyond the dry grass and eventually I noticed specks of green at the base.

'We only stock a third of the country at a time. The rest is spelling.'

'Spelling?'

'It's important not to flog the country. Spelling gives it time to recover.'

'I see.' Did I? I was reluctant to admit the full extent of my ignorance.

'People might look at this land and say it's messy. That's because all too often we have an English country garden view of the world.'

I felt as if I'd been caught out, thinking *it's not what I would call beautiful.*

'In Australia we cleared all the trees then suckers came up so we ploughed and planted. We farmed the land or put cattle out to pasture and we killed all the weeds with chemicals. It may have looked neat and tidy but it wasn't healthy.'

I began to get a feel for what Philip meant, spotting the 'neat and tidy' cleared land and the 'healthier' paddocks with a surprisingly diverse mix of grass and trees.

'We fence to land type and we never pull the trees from one fence line to another,' said Adele. She got out to open a gate and I breathed in the cold winter morning air, allowing my eyes to adjust to the different land types. Pelicans flew in formation overhead.

'Let's poke on,' said Philip, easing his foot off the brake.

'It's good to see so many trees,' I said, cautiously.

'You like a bit of scrub? So do we. Now, trees. Over there you've got belah, brigalow, myrtle and myall, all trees that provide valuable top feed for cattle. When you see myall growing you know it's pretty handy country.'

Philip pointed out box and mighty red gums lining the creek. I was fascinated by his ability to identify so many species that to my untrained eye looked remarkably similar. 'That's cypress pine, over there is wattle and ironbark. The mulga you can see over there is a shallow-rooted tree, rich in protein, and we've got bottle trees, false sandalwood and moolie apples too.'

He went on to explain that there was a time, not so long ago, when the sweet country we were driving through was virtually unusable, blanketed by a thick layer of prickly pear. The highly invasive plant was introduced into Australia from the Americas in the hope of starting a cochineal industry (tiny insects that feed on the pear were harvested to make cochineal, a valuable red dye) but its spread devastated the rural industry in Queensland. It was only when an effective biological control was introduced – cacto-blastis – that the land became usable again. Pockets of prickly pear still existed on Dulacca Downs, a reminder of what many people considered the most invasive weed ever introduced to Australia.

The leisurely drive gave Philip and Adele time to explain how their business developed. While the children had been at school and later university (Lachlan studied agribusiness and Alister opted for marketing) Philip and Adele were free to concentrate on building up the business. One thing they'd insisted on, right from the start, were regular family meetings. Everything was on the table – levels of debt, succession planning, purchasing decisions – to keep the children involved.

'We thought if the kids ever wanted to work in the business, we had to make sure they knew what was going on.'

In the early years there were tax debts to pay, strategic restruc-turing and a lot of deadwood to cut from the operation. Philip's experience with Stanbroke stood them in good stead. Six years later they were trading along well, breeding cattle in the Cape and fattening at Banchory when suddenly, in Philip's words, 'the wheels fell off'.

In 1998, for personal reasons, Philip's brother wanted out of the business. The only option at a time when they were carrying a lot of debt was to split the property at Banchory and sell the most valuable half. Philip and Adele took on more debt to buy his brother out, leaving them with an undeveloped 40,000 acres.

And that's when their road got bumpy. In subsequent years they bought cattle when the market was booming and sold when it was depressed.

'When we couldn't afford to buy any more cattle we had to use the land at Banchory for agistment.'

Surprisingly, in the recesses of my mind, I knew what that meant. Agistment involves leasing land to others so they can graze cattle on it. They pooled 20,000 acres to begin the long process of developing it with fencing, water and grass.

Fluctuations in the dollar had a huge impact on the export market around that time – a strong dollar meant weak export prices – and with over seventy per cent of bullock cattle in Queensland destined for export, Philip and Adele were, in their own words, 'over it'. They held a meeting with their sons, documented their discussions and took the radical step of deciding to align their business with the domestic market.

'I'd always had the idea of paddock to plate in the back of my mind,' said Philip. Looking back, he was forced to admit that he and Adele had very little understanding of how the market worked, and 'we sure left a few dollars along the track finding out!'

There was that big laugh again, hinting at the resilience and philosophical approach to setbacks that had seen them through some incredibly tough times.

As Philip explained it, most of the domestic meat industry in Australia revolved around a handful of players, with big volume and low margins, so once they'd made the decision to enter the domestic market, they had to find a way of competing.

The first stage was to lease a property, borrow money for stock and get back in the game, or as Philip puts it, 'We jumped off the cliff with a bucket of borrowed money and tipped it into the water.'

They looked for a fattening property in the Darling Downs, an area that Philip had noticed was always green when he drove through from southwest Queensland. Which was how they found Dulacca Downs. The property was small – just 8000 acres – but it ticked all the boxes, especially when Philip looked closely at the trees. 'It's always an indication of good country when you see brigalow as a big tree in the same area as belah.'

They set about progressively breeding at Banchory and back grounding (growing) at Dulacca, adding another 5000 acres of land with the purchase of an adjoining property, Heatherlea. They put Angus bulls over Brahman cattle, mixing the heat-resistant *bos indicus* Brahman cattle with the Scottish *bos taurus* Angus, introduced to Australia in 1840, and before long they'd aligned themselves with Australian Country Choice, the principal northern supplier of meat for Coles supermarkets, as well as for a number of export customers.

It sounded like a prestigious position to be in and I suggested as much.

'It was, but we still weren't satisfied. We'd always had an idea in mind that we wanted to get closer to our customers.'

First they tried opening a butcher's shop north of Brisbane. The Vietnamese owner who had been struggling with the shop offered it to them for six months, providing they were willing to stock and manage the meat cabinet. The experience taught them about ordering systems and invoicing, as well as signage and display and the logistics of delivery. They also learnt how to make great sausages (use iced water or the fat won't bind) but they didn't make any money.

Next they tried retail ready into food stores, then they tried supplying farmers' markets and when that didn't work (a lot of fun but not enough volume), they tried setting up an alliance with a portion cutter who ripped them off.

If nothing else, the frustrating setbacks taught them a lot about how the industry worked. In the past, as primary producers, they had always focused on working with cattle to get the product right. 'We'd load the animals onto a truck, pat ourselves on the back and all the time we were just waiting to get screwed.

The mantra for most primary producers is "those mongrels, they've done it again!"'

Now that they could see what was involved in the entire process, those 'mongrels' probably didn't seem so unreasonable.

The twists and turns led finally to the establishment of Rangeland Quality Meats, a paddock-to-plate premium beef production business set up in 2011. At the time of writing, the fledgling RQM had already won one award for excellence, with Philip and Adele named Producers of the Year at the 2012 Queensland Red Meat Awards.

More than ten years had passed since Philip and Adele had thrown their bucket of money into the water, and I was curious to know how well they were doing financially.

Philip gave a big laugh. 'Well, two years ago the bank manager called a meeting with us and he said, "When is enough, enough?"'

He gave a rueful smile as he recalled that meeting.

'So that was it, make or break. We're committed now. I reckon we've paddled back up to the surface for air a few times and now we've found a hollow log we're floating on.'

In the years since they shifted their focus to the domestic market, Lachlan had joined the family business and now lived at Heatherlea with Anna and their young son, William.

It was obvious that Adele and Philip were delighted that Lachlan had chosen to join the business. 'He's always had a way with animals, far more than me,' said Philip. 'My horsepower has had to be directed towards marketing and managing the business in the past couple of years. You have to be logical or you can lose a lot of money very quickly. The margins are small and you can spend $56,000 a week on cattle, $12,000 on boning, plus freight and associated costs. If you take your eye off the ball for a fortnight you can be down a lot of money.'

We found Lachlan quietly tensioning a wire fence with a steel bar, strengthening the corner section where cattle like to sleep. He didn't have his father's height but he resembled him in many ways – both deep thinkers, both passionate about the land. With Philip it was trees; with Lachlan, soil.

'I'm not a conspiracy theorist, but if this farm didn't need any chemical input it wouldn't help anyone else's business. Other businesses want us to use chemicals, then an agronomist to advise on which chemicals to use, and fertilisers to lock up the soil.' Lachlan shook his head. 'We don't need any of it. If you don't flog the land it recovers and regenerates.'

Philip admitted he took a bit of convincing. 'I came from a corporate background, which had a very controlling mentality. The answer to most problems was a chemical blast, although I never felt easy about that. Since Lachlan joined us we've travelled a long way down the path towards organic solutions.' The only thing that seems to have stopped them from applying for organic accreditation was an ongoing tick problem in the north, where they still breed.

'I know that the land will provide for us as long as we look after it,' said Lachlan. 'And the more I try to achieve a balance between production and sustainability, the more convinced I become that this is the model for sustainable farming.'

We drove on and Philip explained that it was now Lachlan's responsibility to oversee operations at Dulacca Downs and Heatherlea, where cattle were brought in at around 280 kilograms and held for six months.

'During that time they're moved from paddock to paddock, allowing time to spell the land, until at around 360 kilograms Lachlan walks them over to an adjoining feedlot to be grain-fed for fifty or sixty days. Low-stress handling is a big part of what we do. High stress affects the quality of the meat and every week fifty-six head of cattle have to be ready to send, all with full trace-ability from paddock to plate.'

The Hughes' entire process, from initial breeding to low-stress stock handling to grazing on good pasture to grain feeding, was set up to ensure a consistent supply of top quality, tender meat.

I had wondered how I might feel seeing animals close up whose ultimate fate was to be slaughtered for meat, and the answer was, a little bit sad. But I've never been a radical vegetarian campaigning to stop the slaughter of animals; vegetarianism was simply a personal choice I made many years ago. I would never stop others from eating meat.

More importantly, I was relieved to see that the cattle in question seemed docile and inquisitive, relaxed, well cared for and well fed.

Producers like Lachlan and Philip cared about their animals.

*

'Where's the butter, Mum?'

'In the fridge.'

'That's margarine.'

'I know but it's spreadable.'

'Mum, what are you thinking? Look at all those acidity regulators!'

'Then stop reading the packet,' Adele said with a throaty chuckle.

We were preparing dinner on my last night in the kitchen at Dulacca Downs, and it was clear that Hughes family relationships were marked by abundant goodwill and high-spirited good humour.

'Seriously, Mum, fructose is really bad for you,' Lachlan insisted.

'We're all going to die!' intoned Philip, and Adele let out another laugh.

Lachlan was clearly used to the ribbing and he pressed on regardless. 'Disease due to lack of minerals in so-called "healthy choice" frozen meals will get you long before any food will kill you.'

'Another glass of wine, Anna?' Philip asked, innocently.

Much like her mother-in-law, Anna was a city girl from Brisbane who'd had a hankering for the bush. A high achiever who excelled at music and sport, she drove herself hard, piled on the pressure in leadership positions and peer mediation programs, then gave it all up to head west and work as a jill-aroo in Haddon Corner. 'I hated the city, I always felt lost there. My great grandfather was a saddler, so maybe there was some genetic imprint that drew me to the land.'

The talk turned to soil, to what might underpin the ten-kilo weight difference between cattle at Dulacca Downs and Banchory.

'Whatever feed grows, cattle get fat on it,' said Philip, 'but after a run of, say, three good seasons with long grass the cattle won't get any fatter. So are there limited nutrients, no matter how high it grows? Maybe the country sacrifices what grows on top to get nutrients back into the soil?'

'Corn syrup! That's really bad as well, did you know that?'

'Is this fish done?'

The multiple conversations were fuelled by wine and laughter. I hesitated to ask a direct question and break the mirth. But I was keen to know how Lachlan felt about coming back to work alongside his father. I'd already heard from Adele, who admitted he'd found it hard after all those years at boarding school and university to come home and have his mum tell him to pick up the towels from the bathroom floor. Moving to an adjoining prop-erty would have helped but what about his father questioning the logic of theories he'd been taught at university? Philip had already admitted it hadn't been easy, and that he'd had to 'pull his head in'. What about Lachlan, I wondered? How difficult had it been for him? Had he ever thought of quitting?

'It was never going to be easy. I was very well looked after when I was young, then I went away and when I came back I had to be an individual, but I was with family who knew me long before I knew myself.'

A deep thinker then, like his father.

'You can do any amount of workshops on succession but the bottom line is, if you don't get on with your dad normally, don't try to do it. If you don't get on, it won't work. A lot of kids just leave,' he added.

There was a slight hesitation, a gap in the air that felt like it carried the memory of all the disagreements they must have had, all the issues they'd worked through, the battles of wills and the clashes of ideas.

'A lot of fathers are narrower than your dad,' said Adele, loyally.

'He had a few wins and I had to listen,' Philip admitted. 'What's the point of having someone who is university-educated if you're not going to listen to their ideas?'

Lachlan must have been helped by his father's forty years of experience and his willingness to step back and let his son get on with the job.

'Fundamentally you've got to get to a place where you trust each other,' Philip said.

Lachlan nodded in agreement and the hesitation this time was barely noticeable.

'We just look at what's wrong and we find a way of making it work.'

Before I left, Adele and Philip drove me out to Heatherlea, where Lachlan and Anna's self-sufficient life bore a remarkable similarity to the life described by Philip and Adele on some of the big corporate properties they worked on in the early days of their marriage.

'It's a lot of work but good luck to them,' said Philip. 'We loved it.'

There were chooks, ducks, cats, dogs and guinea fowl that roosted in the tipuana tree at night and warned of approaching snakes. Lachlan bred horses and Anna bred kelpies, with several sheep for meat and to use as a miniature 'flock' to train her young dogs.

'And we're thinking of getting a milking cow,' she added.

Anna shared her husband's passion for all things organic, avoiding chemicals in household cleaners, refusing to buy genetically modified products and shopping for unpasteurised, non-homogenised milk.

'If we run out of rainwater for drinking we use filtered dam water.'

I wondered how safe that might be.

'It's fantastic! There's not a lot of cattle churning it up and I'd rather use that than town water. Last time we ran low we got town water delivered by tanker and it killed my sourdough culture!'

I followed Anna to the back of the house, where a burgeoning vegetable plot was taking shape.

'We grew everything organically from seed last year, planted them out at the right time according to the moon, the sun and the stars – Aphrodite, you name it – then along came summer and bam! It all died. Summer smashed us.'

Like other members of the Hughes family, Anna had a great ability to laugh at herself.

She was standing under the shadecloth she'd erected to cover the organic vegetable plot that was such a failure in its first year.

It now boasted a flourishing crop of onions, tomatoes, snow peas, leeks, broccoli, silverbeet and herbs.

'Hey, you need a bit of poison in there,' said Philip as he appeared around the corner.

Anna refused to take the bait. She struck me as a feisty, independent woman who was a good foil for the more reserved Lachlan.

'Darl, if you see the hose on, and I've been away a while, can you turn it off?' Lachlan asked her politely.

'Yeah right, because I've got nothing else to do, I put my knitting away as soon as I saw you coming.'

He smiled.

'Love you, darl.'

'Love you too.'

*

On the drive back to Roma airport, Philip and Adele told me about the latest development to thrill them both. After several years working in marketing with Red Bull, their youngest son, Alister, had decided to put his marketing degree to use with RQM.

'We've learnt a lot,' said Philip, 'and we're in this business because we believe in it. If the boys can understand how business works, they can use that knowledge anywhere. Mind you, it's a slow process and there's always someone over the hill making all the money, doesn't matter where you are or what business you're in.'

He laughed, and his face lit up. 'And after all that we still haven't made any money! Yeah, you'd have to wonder about us, wouldn't you?'

Virginia and Steve Chilcott

'Glasslough'
Epping, 40 kilometres southeast of
Launceston, Tasmania

'Hazel Park'
Meander Valley, 60 kilometres southwest
of Launceston

I was keen to interview Virginia and Steve Chilcott for several reasons. They were a couple starting out on the journey of raising a young family, they were both dedicated farmers and they had an unusual set-up. The commodities Steve and Virginia farm are vastly different from each other – Steve is dairy, Virginia is sheep and cropping – and traditional farming wisdom would tell you that such diverse operations are chalk and cheese.

The couple live in Tasmania's Meander Valley, high up at the base of the Great Western Tiers, with their two young children: Henry, born in 2011, and Georgie, born two years later. Farming has been in their respective families for generations.

Since their individual farms are an hour's drive away from each other – Steve's dairy is in the foothills of Mother Cummings Peak and Virginia's sheep farm is at Epping Forest, on the sunny plains forty kilometres southeast of Launceston – they faced extreme demands on their relationship in the early days. Both driven to succeed, neither of them thought twice about working through the night. Somehow, though, they had to carve out time for each other; the answer was to commute.

At the end of a long day on the farm, one of them would shower, change and drive a hundred kilometres so they could spend the night together at the other's farm. It all depended on who was busiest. Next morning, hours before dawn, and long before most of us would dream of setting an alarm, that same person would rise and drive back, ready to start milking, crutching, lamb marking, sewing, harvesting or calving – or any one of the hundreds of other jobs that constantly need doing on a farm.

The arrival of children added to the demands, yet somehow they still managed to run two businesses, shuttling between both. Virginia even found time and energy to compete at the highest level in the national finals of the yard dog championships. They also found points at which their businesses could be mutually supportive, revealing an unexpected synchronicity that allowed them both to continue farming, and to raise a family.

Inevitably there have been compromises, one of which was the choice of where to establish a permanent home. The inescapable daily demands of dairy farming made Meander the obvious choice, so Virginia commutes to Epping Forest most days, dropping Henry off at day care before continuing on with Georgie, who has probably spent more time in a paddock than most babies spend in a pram.

I visited Virginia and Steve in mid-winter, on a night of bitterly cold rain with frost and snow forecast for the next day. It was the start of the calving season for Steve, a period of relentless demands on his time and sometimes twice-nightly visits to the paddock.

It struck me that Virginia and Steve's characters reflected the nature of their different farms: Steve more placid, calm, solidly built and ruminative by nature; Virginia the chattier of the two, more outgoing and cheekier. Both, though, were highly successful farmers in their own right.

Chalk and cheese

Rain fell in horizontal sheets across paddocks small enough to see from one side to the other, some of them bordered by familiar-looking hedgerows. I could have been in England. I was instead in Tasmania, heading for a farm less than an hour away from Launceston airport, on a cold winter's night. Sat nav had failed me yet again and I was hopelessly lost.

I'd been a bit blasé about visiting Tasmania. The whole island was only 286 kilometres long and I had the luxury of a street address. How hard could it be?

Damn near impossible. Sat nav refused to accept the house number I'd repeatedly tried to enter so I chose a random number and set off for the Meander Valley, peering through driving rain at occasional farmhouses set well back from the road. There were no streetlights (I know, it was stupid to have expected them) and more often than not the farm entrance gates had names not numbers. To make matters worse, high mountains blocked the mobile signal so there was no hope of calling for help.

Suddenly sat nav sprang to life. *You have arrived at your destination.*

Really? I was on a dirt road in the middle of nowhere with only cows for company. One of them peered through a nearby hedge, rain dripping from its ears, steam rising from its nostrils. I drove on, cursing the whole ridiculous notion of travelling around Australia to interview people for a book on farming families when I wasn't a farmer and I had no children (I also had no sense of direction).

Admitting defeat, I drove through a farm gate and sat in darkness, listening to large dogs bark while I waited in vain for someone to come and find out why. Finally, thankfully, the phone locked on to a signal and I rang Virginia for directions.

The sight of her smiling face peering through a window was so welcome I could barely stop myself from flinging open the car door and rushing across to hug her. Perhaps I did, I don't remember now.

'Steve's just heading out to check on a cow, would you like to go with him?'

It was ten-thirty at night. The temperature outside was barely above freezing and an enormous log fire was burning in their living room. A cat lay asleep on the arm of a squashy sofa and in that precise moment I knew I could never be a dairy farmer.

'I'd love to,' I said.

Steve scanned the paddock, humming quietly under his breath, using a powerful beam of light mounted on his truck. The light fell on a large mob of docile cows that all turned to stare. There was no let-up in the cold rain that had been falling steadily all day and Steve was searching for a cow that should have given birth earlier.

'Heifers can get a bit silly.'

'Why?'

'They've never given birth before.'

We got out of the car for a closer look, gumboots squelching on wet grass as we approached a group of black and white cows. Steve was a tall, powerfully built man yet the cows seemed to dwarf him. Had cows always been this big? The Friesians I remembered from my childhood (we grew up on the edge of a Gloucestershire village surrounded by fields full of dairy cows) seemed smaller. Perhaps memory was playing tricks on me, or maybe they bred them bigger in Tasmania.

'She should have had it by now. On cold wet nights they sometimes tuck up against the fence and they won't even try to calve.' He sounded worried. So was I. How did this not qualify as a cold wet night?

Unseen in the darkness behind us was Mother Cummings Peak, one of the highest peaks in the Great Western Tiers. The paddock we stood in sloped gently away from the road, what Steve called a 'run-off' block, where cows were brought to give birth. The nights could drop to minus five in the Meander Valley and daytime temperatures sometimes struggled to rise above freezing. Steve seemed impervious to the cold. His focus was on a large beast standing alone, the protruding hooves of an unborn calf clearly visible.

He walked quietly up to the heifer, humming all the while. She shied away and other beasts crowded in. They weren't threatening, just curious. Steve advanced on the heifer, using his calm presence to lead her into a fenced-off area under the trees, away from the rest of the herd. Forgetting the log fire I'd been longing for, I followed. I'd never seen a calf born before.

'Wa-ay go, wa-ay go,' he murmured softly, gently reassuring the nervous creature as he manoeuvred her into the yard. Once she was safely penned, Steve tied a rope around the protruding hooves. Using his not inconsiderable strength, he then leant back and pulled, hard.

'Come on, girl, come on.'

The heifer emitted a low call and Steve pulled on the rope then released, pulled and released. The call rose to a crescendo of sudden noise and a calf slipped out, hitting the ground with a wet thud. I gave an involuntary sigh. The cow nudged her calf, licking and sniffing it. It didn't move.

'Stillborn.'

The elation I'd felt was replaced with sadness and a sense of shock. In place of the anticipated newborn baby calf, which only seconds before I had imagined struggling to its feet, was a slippery carcass lying on the ground. Steve had no explanation for the stillbirth.

'Over a season we might get fifteen or so. It happens.'

His pragmatism was born of twenty-five years of experience and his focus switched to the heifer that had just given birth.

'The main thing now is to look after her.'

After checking the heifer would be all right, then scanning the others to see if any of them were likely to give birth later, we drove back to the inviting warmth of their modest, single-storey house, heated by that huge log burner that stayed alight night and day over winter.

My romantic view of farming took a bit of a knock that night, forcibly reminding me that this was a business with no place for sentimentality; not that Steve didn't care for his animals. In a herd of 350 cows he could recognise all of them as individuals.

'I've been working with dairy cows since I was fifteen. It's all I ever wanted to do,' he explained. Around fifty-five cows would be calving in the next six or seven days, marking the start of what would be a frenetic three-month period of calving for most of the herd. Steve was well aware that over that time Vee (as Virginia liked to be called) would virtually be a single mum. He was full of admiration for his wife's ability to juggle farm work with motherhood.

'She's a really good mum and a really good farmer,' he said simply. Steve was a man of few words, his self-effacing strength a marked contrast to Vee's energetic confidence. She was quick to return the compliment though.

'Steve's too modest to make a big thing about it, but he took a run-of-the-mill dairy farm and he punched it into the next generation. It's had massive productivity gains since he took over.'

They met relatively recently, in 2008, when Vee was in her late twenties and Steve in his late thirties. The long drought had just broken in Tasmania and the first decent rain for five years brought with it an explosion of wildlife. Steve had joined a group of shooters on wallaby control at Vee's sheep farm in Epping Forest, and she was immediately taken with his quiet strength.

'He was a big bloke, proud and tall, and I teased him about needing such a big gun. I suggested it might be compensating for something else.'

Far from taking offence at Vee's ability to overstep the mark, Steve asked one of the other shooters for her phone number.

Vee wasn't sure. Dairy farming was messy and high-input, sheep were tidy and low-input. 'I didn't think I could have anything in common with someone who worked ridiculous hours and milked cows that stank.'

She was wrong, of course, and after a flurry of text messages they started dating. A shared love of outdoor pursuits often led them roo-shooting and Vee took great delight in telling me how Steve once accidentally shot her in the buttocks.

'It was a love bite.' She laughed.

If someone had shot me on a date I think I'd have called the police.

'Of course, if we'd met in winter, at the height of the dairy calving season, I probably wouldn't have given Steve a second look. I would have thought he was mad. Dairy farmers work all hours of the day and night during calving. What's the point in having a relationship with someone you never see?'

Both Virginia and Steve were used to working exceptionally long hours, investing all their time and money into growing their respective farms, and they had both spent time in unsuccessful relationships before. Neither wanted to repeat the mistakes of the past. The answer was to divide their time between the two farms, taking turns to overnight. Whoever didn't drive did the cooking.

Six months into the relationship the crazy winter period hit and Vee was crutching, lamb marking and vaccinating her crossbred ewes until late at night. Steve would overnight at her farm and be gone by four-thirty in the morning. When calving started Vee would overnight at Steve's farm, catching what time she could with him when he wasn't spending long midnight hours in a cold wet paddock.

'There was a whole lot of hard-mettle testing in the relationship. If we weren't going to make it we would have discovered very early on.'

Vee never thought she would get married – let alone marry a dairy farmer – but she was forced to admit there was something special about Steve. On their first proper holiday together in New Zealand, she kept waiting for him to propose.

'I knew he was the one, he knew I was the one and time was ticking for us both. I kept thinking *when* is he going to do it?'

Vee's eagerness for him to propose at the top of Mount Remarkable was partly to do with the idea of a return ski trip to celebrate their anniversary each year, but Steve had other ideas. He proposed when they got back to Meander, in a remote camping spot where they'd enjoyed their first picnic together, near the waterfall that marks the start of the Meander River, high up in the Western Tiers.

I wondered if Vee had changed her opinion on the relative merits of sheep versus dairy now that she was married to a dairy farmer?

She shook her head. 'Sheep may not be as profitable but dairy cows produce runny smelly crap that can land on your head.' She paused. 'Fortunately Steve is bald, so it washes off easily.'

Vee slipped into the room next door to settle their five-month-old Georgie, and Steve padded in to check on two-year-old Henry, who was quietly murmuring in his sleep. Over the next twelve weeks they would have two needy children inside and several hundred outside.

Early next morning frost lay thick on the ground and mist hung over the valley, shrouding the winter-bare fruit trees at the front of the house in a cloak of white. The eerie call of peacocks, heard but not seen on such a morning, echoed across the valley as Henry finished his toast and slipped out of the house, jumping off the veranda and heading for the nearest frozen puddle. I wondered if he should be doing that in just his socks and pyjamas.

'Henry, get back in here!'

Henry reluctantly stomped back to the house.

'It was so much easier before children came along,' Vee said with a sigh, sitting Henry on a chair and pulling off his muddy socks. Henry reached for her computer.

'No, Henry.'

He picked up her phone and started punching numbers to make it beep.

'Henry, no, that's mummy's phone.'

Henry kept punching.

'They're gorgeous children really, we're so lucky. I only some-times think about sending them back. Henry's going quite cheaply right now,' she added, lifting Georgie out of the rocking cradle for a morning feed. Steve had already left to check on his cows.

From what I'd heard, Vee was an extremely capable farmer who had invested considerable time and money building up a business that was now a hundred kilometres away from where they lived. I wondered how meeting someone with an equally demanding business had affected her. When I was younger I frequently put my own needs aside to focus on a new rela-tionship. Had she done likewise? And how did she juggle the demands of motherhood and farming?

'I was used to working from dawn to dusk. When you've only got yourself to please you can be hugely productive.'

At twenty-three, long before she met Steve, Vee had inherited part of a family farm that needed hundreds of thousands of dollars' worth of investment. She took out a sizeable loan, knuckled down to work and took off-farm jobs as a wool classer to bring in extra cash and set up a lamb feedlot for a restaurant owner. She thrived on the challenge of seeing how late she could work and thought nothing of finishing crutching sheep at one-thirty in the morning. In a tiny weatherboard cottage she lived like a pauper and paid off the loan within three years. It was for $160,000.

Vee was obviously an extremely capable farmer. As soon as the last payment hit the bank she applied for another loan. 'And this time I didn't have to swallow my pride and ask Dad to stand as guarantor.'

The tension in her voice suggested she didn't have the easiest relationship with her father.

'Okay, Henry, time to go, where's your coat?'

The mist had lifted, revealing mountain slopes that rose behind the house and a breathtaking view across the valley. The peacocks I'd heard earlier were Steve and Vee's, so were several dogs chained up at the back. Apart from the peacocks (which for some reason I equate with stately homes in England) it was a modest, understated picture with a large machinery shed next to the house and a trampoline on the front lawn.

Vee buckled Georgie into the back of the car, rescued Henry from another puddle, and we set off for day care in Deloraine, en route to her farm in Epping Forest.

'I honestly never thought I'd get married. There aren't many men who like strong, opinionated women.'

I wondered how many male farmers would have described themselves as opinionated? Virginia was a strong, self-assured woman whose confidence was well placed. It was also undercut with humour. At one point she said, 'I hope you won't describe me as short and fat.' (I won't because she was neither.) 'Georgie is fairly fresh, only five months. I've found it hard to lose the excess. Too strong a liking for Kit Kats.'

A family friend had described Vee as someone with 'lots of go and a great personality, a lot of fun'. It was an accurate description. She was entertaining company on the hour-long drive to Epping Forest – a drive she did almost every day – and I listened with growing interest as she told me something about her extraordinary family background. Much of it centred on her grandmother.

Vee's paternal grandmother, Joan Prevost, was born in Sydney in April 1914. She met her future husband Edward (Ted) on a trip to the UK in 1936. Ted was an officer in the British navy and Commander of HMS *Hood* during its sea trials (although not when it was sunk by the *Bismarck*). He was also fourteen years older than 21-year-old Joan.

At the end of World War II Ted was decommissioned and Joan pushed for them to move to Australia. She thought they would have more opportunities in the land of her birth. After a short spell in Sydney they settled in Tasmania, where they bought a sheep property.

Joan had no background in farming and Ted's connection with the land was tenuous. He'd been raised on a farm but sent away to school at the age of seven then into the navy at thirteen, so neither of them knew the first thing about farming when they bought Glasslough, a property built in the 1800s near Epping Forest on the banks of the South Esk River.

'I'm sure she would have been urging Ted on, telling him they could do it.'

Joan Prevost discovered she had a natural ability with sheep. That ability was coupled with a ferocious work ethic that led to the highest accolade possible at the sixty-seventh Melbourne show. In 1949 one of her Polwarth rams was awarded the championship prize. For someone who had only bought her first mob of sheep three years earlier, it was an extraordinary achievement.

Joan and Ted renovated the rundown estate and built a thriving business, directed largely by Joan who became a well-respected farmer in her own right. Vee had the greatest respect for her. 'She was an amazing woman, still sharp as a tack and trading shares right up until five years ago. I remember growing up she was the matriarch of the family, there wasn't much wriggle room . . .'

Vee had a lot in common with her grandmother, although she hated it when her siblings said she resembled her.

'What child wants to be told they look like a seventy-year-old?'

She could see now how flattering those comparisons were. 'If my grandmother saw an open door she would walk through it. I would love to be able to do what she did, which was to set up the next two generations so they were financially secure. She left an amazing legacy.'

Joan Prevost died not long after my interview with Vee in 2013, at the age of ninety-nine. Alison Andrews' obituary of her in the *Launceston Examiner* on 24 October 2013 was fulsome in its praise. 'Joan Prevost never had time for the conventional wisdom on the roles of men and women . . . and nobody ever had the courage to point out to the canny Northern Tasmanian sheep farmer and Australian stockmarket expert that she was often a rare woman operating in a man's world.'

Joan Prevost had clearly passed on her genes to her granddaughter.

Vee was the youngest of three children, five years younger than her brother James and seven years younger than her sister, Katie.

Five years seemed like a big gap between siblings and Vee didn't surprise me when she admitted she didn't have much sense of connection with her brother and sister when she was growing up. I was more surprised to hear that she didn't have much of a connection with her parents either.

'I don't think my parents really got me. They didn't pick up how sensitive I was and how much I cared about the welfare of animals.'

When anything upset her, Vee turned to animals for comfort; it wasn't unusual for her parents to find her asleep at the back door, curled up with the dogs on their mat.

'I hated school, especially in years eight to twelve when I boarded because it took me away from my animals. All I wanted to do was go home and work on the farm.'

Vee's caring nature and strong connection with animals at one point led to a four-year period as a vegetarian – unusual behaviour in a sheep farmer's daughter who had since gone on to become a successful farmer in her own right.

'I had more in common with animals than with people. I wasn't an intellectual child and it was easier to relate to dogs and sheep than it was to talk to people. I didn't know how to sit down and talk normally about my feelings. It took me a while to learn that death was part of the cycle of life.' When I heard the story of what happened to her sheep, I understood why.

Strays and orphans always found a champion in Vee, who appointed herself chief feeder of any animals needing extra care. By the age of six she had accumulated a mob of pet sheep, all of which were led through the garden to feed on a daily diet of apples from the orchard, roses, flowers and low-hanging leaves. If leaves were out of reach, Vee would pull down branches so the sheep could stand on their hind legs to feed. To vary their diet, she would open the silo chutes and give them a line of grain.

Like the Pied Piper, a single long whistle from Vee would bring her pet sheep running. She would lead them into another paddock, only to get bored then lead them back again moments later. She had names for every one of them.

When her father, Rick, counted 200 in his daughter's mob one day he decided enough was enough. While Vee was at school he rounded up the sheep, including her pet sheep, and he sent them all to the shearing shed. When shearing was over he incorporated her pet sheep into the main flock, making sure they were with the right 'mobs'.

Vee got off the school bus at five o'clock that afternoon and stared at the empty paddock near the tennis courts.

'Where are my sheep?'

'Oh, they're around here somewhere,' her father answered, nonchalantly.

Vee marched down the drive (which when I saw it must have been at least half a kilometre long) and stomped off towards the hill paddock, where she gave a long whistle. Every single one of her sheep lifted their heads. Then they galloped down the hill onto the main road and trotted up the driveway with her.

'Don't do that again,' she muttered angrily, pocketing a notebook she carried that tracked the progress of her mob.

Like most farmers Rick Prevost had a killing paddock, and when the farm needed meat he would catch and kill one of the sheep. His sensitive daughter struggled to accept the stark reality that an animal had to die in order for her to enjoy meat. Vee's strong connection with animals made it deeply upsetting for her when she witnessed the killing. It was bad enough when she didn't know the sheep, but one day Rick did something that sounded unthinkable to me.

Vee was about ten years old. She spotted her father walking through the paddock where her pet sheep were grazing. She saw him send one of his dogs around the mob and she ran down to see what was happening. That's when she saw the knife in his hand. Rick pointed at a wether and told his daughter it would have to be killed. It had lice, he said by way of explanation.

'No, it doesn't. I know it doesn't!'

Vee sensed that her father just wanted meat for the house and her pet sheep were easy to reach. She yelled and pleaded with him, she even tried shouting at the dog to send it away, but none of it made any difference. Vee watched in growing horror as her father grabbed the sheep.

'Don't!' she screamed. He turned the sheep over and slit its throat.

Vee was devastated.

'I couldn't believe he'd done it. If Dad had killed a wild sheep I still would have been upset, but to kill one of my pet sheep in front of me was awful.'

Her family viewed her reaction as over-emotional and Vee withdrew further into her own world. She spent the next four years as a vegetarian and it was only when she left Tasmania to work in America that she learnt to accept the cycle of life and death.

As a vegetarian – and an animal lover – I was stunned by Vee's story. Of course I'd only heard her side of it and, yes, you had to kill an animal if you wanted to eat meat, but how could any parent have failed to recognise how sensitive she was? Had her father been trying to teach her a lesson? *We live on a farm, we eat meat, why be sentimental about killing?* If so, it was a brutal lesson. And where was her mother in all of this?

We dropped Henry off at day care, stopped briefly so Vee could feed Georgie, and continued on our way to Epping Forest, where Vee's parents still lived. After her story, I wasn't at all sure how I felt about seeing them.

I was glad my first meeting with Vee had been on the mountain slopes of Meander in the modest house that she shared with Steve and their two young children. I think I might have been overawed if we'd met at Glasslough.

It was an absurdly picturesque setting. The turn-off from the main road led down a great sweep of productive farmland that ran for several kilometres, dipping into a fertile valley towards a substantial two-storey house perched high above the banks of the South Esk River. This surely had to be what farmers called 'sweet country'. A broad stretch of water curved invitingly in front of the gracious house, set against a backdrop of the Western Tiers and Ben Lomond.

Vee pointed out the Angus cattle and poppies that her brother farmed on one side of the access road and the lush green oats she had planted on the other.

'See that centre-pivot irrigator?'

She pointed to a vast piece of machinery, like a metal praying mantis, in the centre of one of the paddocks.

'When I took over the farm it was the first thing I bought.'

The centre-pivot irrigator had cost Vee $156,000.

'How old were you when you took over the farm?'

'Twenty-three.'

Vee's grandmother had built up the business and she'd handed it over to her son, who in turn handed it over to Vee. Had it been that simple? Did the family calmly negotiate a succession plan and did Vee wait patiently for power to be handed to her? Or had she seized it in a classic Shakespearean scenario so familiar to politicians? Twenty-three sounded young to be running what looked like a substantial operation, but as someone in her mid-fifties perhaps I'd forgotten (more likely I never knew) what ambition felt like. Young people could be hungry to succeed and Vee was clearly ambitious.

While she was studying agriculture at Dookie College in Victoria Vee found a novel way of funding some of her living expenses. Having persuaded the college to give her a small mob of training sheep she then used them to retrain wayward dogs. Some were pets with too much energy for their previous owners to handle, others had simply been abandoned.

When she wasn't studying (or partying) Vee spent all of her spare time in the paddock with her mob of sheep, retraining an unruly collection of dogs that would otherwise have been forcibly put down. Sometimes after only weeks in her care, a dog's errant behaviour would be channelled into work and it could then be sold to a willing farmer. The profits helped fund her college education.

Vee went back to the family farm in Tasmania as an eager twenty-year-old, brimming with enthusiasm and ideas, and in typical fashion she immediately proposed changes to the way the farm operated.

'I came back and I just wanted them to get out of my way and let me do it.'

I began to feel a twinge of sympathy for her father. It sounded like he'd been squeezed between a fearsome matriarch of a mother and a forthright daughter. Vee's grandmother had retained a close involvement in the running of the farm and she and her son Rick ran Merino sheep for many years. Tasmania is one of the few places in the world that can successfully breed Merinos (I hope any sheep farmers out there are nodding) and Joan and Rick made the most of the opportunity with high stocking rates. Vee had other ideas.

'Dad could never grow enough feed. He had the philosophy of growing the grain and hay to feed the sheep, and that had merit in some situations but not ours.'

Rick had followed in his mother's footsteps, producing fodder in spring and summer to feed the sheep throughout winter. With the benefit of her agricultural training, Vee was convinced that wasn't necessary.

'He would have made more money from selling the commodity, not putting it down a low-profitable sheep's throat!'

Vee wanted to grow wheat to sell and let the sheep graze on grass. She pushed for quality, not quantity, believing a smaller number of fat, high-producing crossbreeds were preferable to 'starving Merinos'. Apparently the less Merino sheep eat, the finer their wool – no wonder animal-loving Vee didn't approve.

'I reduced the flock by half. Being a larger-framed animal that consumed more grass, it was more like three quarters of the original Merino operation.'

Vee met stiff resistance to her ideas from her father and grand-mother, but she pushed on regardless. She started mending and replacing fences and looking at ways to improve productivity from the livestock. Irrigation was another bone of contention.

'Dad used a soft-hose watering system.'

The tone of her voice made it abundantly clear exactly what she thought of such a system.

When it dawned on Vee that she was improving an asset that would ultimately be split three ways she put the brakes on. Before investing any more time and energy into the family business she tackled the thorny issue of succession head-on.

I could imagine how heated those discussions must have been. Vee joined an external farming advisory group that allowed her to bounce ideas off other people without the emotional burden of family ties.

'I was pushy, I made succession happen.' (Aha, seizure of power then.)

Vee became convinced that in any succession agreement her father should step aside and cease active involvement in the farm. Her older siblings agreed. Katie worked in broadcasting and had no desire to farm and James was as keen as Vee to sort out a succession plan. Interestingly, it was Vee, the youngest by five years, who took the initiative and made it happen. It took two years of negotiating for Rick to agree.

In February 2005, when Vee was twenty-three, the Epping Forest property was officially split down the middle. One half went to Vee, the other half went to her brother, and off-farm investments ensured an equitable share for Katie.

And suddenly there he was, Rick Prevost, bouncing towards us across the paddock. Any reluctance I'd felt at the prospect of meeting him evaporated because, of course, he was charming, a dapper man with a firm handshake and easy conversation, laughingly admitting in the first couple of minutes to a triple heart bypass operation. I really should have gone through the whole 'I'm not a farmer and I know nothing about farming' routine but I felt slightly awkward and tongue-tied as he shook my hand and started asking questions about farming in England.

Vee came to my rescue. 'Yes, Dad, they do calve indoors in England.'

Rick chatted sociably about lambing and Merinos before finally admitting the real reason he was there.

'Uh . . . there seems to be a mob of sheep . . . not sure why . . . Don't worry I'll get them.'

We watched him drive off and Vee muttered crossly as we climbed back into her Toyota.

'I bet he forgot to shut the gate. I won't jump the gun but I bet that's what happened and now I'll have 900 sheep on my newly planted wheat. Oh that is *so* annoying.'

We sat in the car while Vee breastfed Georgie. Alarmingly, out of the corner of my eye, I saw her brandishing an enormous knife.

'What's this then?' she cooed, cutting a tangle out her baby's hair. It was like a scene from *Crocodile Dundee*.

Then she laughed. 'I know I shouldn't complain. Dad draws on years of experience and there aren't many situations that can't be solved when you add his leverage, ingenuity and sheer cunning!'

*

A log fire crackled and spat in the hearth of the elegant drawing room at Glasslough as Rick stood with his back to the fire, and his wife Leonie sat in an armchair. Curly hair floated in a cloud around her still-beautiful face and I wasn't surprised to learn that she'd once done some modelling for Myer.

'Virginia was my third child, very easy, very good-natured. The others always said she was a mistake but she wasn't.'

'I know I was,' Vee countered quietly, her attention focused on Georgie who lay on the floor at her feet.

'At one point you thought you were a dog, didn't you, dear?' said Leonie, unfazed by the interruption.

Rick laughed. 'Do you remember that time she travelled in the back of the ute with the dogs? It was at the bottom of Goat Hill. They jumped off and so did she! I looked in the rear-view mirror and saw this stunned-mullet look on her face. I could have run her over.'

'I left enough clearance.'

'You were a very particular child,' continued Rick. 'You take after your grandmother.'

Leonie looked faintly amused. 'She was a bit soft about killing sheep. Then she went to America and came back shooting everything. You turned out all right though, didn't you, dear?'

I felt like I'd stepped into a family drama, a film set peopled with minor characters from an episode of *Downton Abbey*. As the conversation progressed I revised my opinion: major characters.

Early photographs of Vee's parents revealed him to be a suave young blade in a white suit and she a willowy beauty. There was nothing to suggest the farmer – or the farmer's wife – in any of the shots I saw. Leonie was a university-educated intellectual who worked for a time at an all-girls Catholic school, St Finbar's. 'They didn't ask me about any teaching qualifications. I was a success the first day because I wore a long purple midi skirt.'

Leonie certainly had no leanings towards farming. Not long after their engagement she and Rick came back from a ball late one night to find Rick's mother in a paddock trying to pull a calf. Leonie was asked to hold the torch while Rick stepped in to help and it was only when the light pitched up that they realised Leonie had fainted.

'They had to pick me up in my ball gown and drag me out of the way. Farming was such a shock. I certainly didn't want to stay at home and be a farmer's wife.'

Leonie spoke with a sweet smile and an air of distraction as she revealed that cooking wasn't her strong point either. 'My mother wasn't interested in domesticity. She taught me nothing. I served peas and mashed potatoes for two weeks when we first got married and when Rick finally plucked up the courage to ask if we could try something different I served carrots and mashed potato for the next two weeks.'

Leonie once tried renovating plasterwork in the house by chipping at it with a knife from the family silver. 'Well it was in the kitchen drawer,' she said by way of defence. 'We didn't ever keep our silver in the kitchen.'

Leonie and Rick took great delight in regaling me with stories about the eccentric relatives who cropped up on both sides of the Prevost family tree.

Leonie's mother left behind a privileged upbringing on a property at Collarenebri to run away to Sydney, where she worked in a munitions factory before marrying the unconventional naval commander, Arnold Holbrook Green. Green's reputation for bravery in battle was matched by his fearsome lack of respect for authority.

He once commandeered a bus in Hobart, ordering the startled passengers to disembark so a group of visiting American sailors could take a tour of Mount Wellington. 'Who's going to pay for this?' asked the amazed bus driver. 'Send the account to the premier,' Green said, dismissively.

Several weeks later the account arrived in the premier's office and Green was called in to explain his actions.

'Only did what you would have done, sir. The reputation of the state was at stake.'

'Oh, well, keep up the good work, Green, well done.'

Rick's father was also a naval commander and a stickler for routine; for fifty years he sat down to Sunday lunch wearing the same jacket and tie he got married in.

'He also wore the same pair of shoes.'

It all sounded so improbable, but some quick research when I got back showed much of it to be true – or at least true enough to have become the stuff of internet legend.

Rick's sister, Dyranda, was a renowned photographer who worked behind the iron curtain in Russia and narrowly avoided being classified as a spy; his brother was a fearless child who roamed the streets of Sydney from a young age, crawling through stormwater drains to see what was in them. The stories went on and on.

With such eccentricity on both sides of their families, Rick and Leonie seemed destined to meet. Born five years apart in the same Sydney hospital, the same matron oversaw both births. At one point their parents lived in apartment blocks directly opposite each other but the couple didn't meet until they were adults, by which time their respective families had moved to Tasmania and Rick was living at Glasslough.

'We met at the races. I think it was at the races, wasn't it? And I remember,' Leonie added, 'seeing your back one night at a ball.'

'Hah! That was me running away.'

They reminisced about 'damp down' days, when as many as 150 waterskiers would be towed along the wide stretch of the South Esk River in front of the house. 'Damp down' days traditionally followed 'dust up' balls.

'Do you remember Paul Vincent swinging from the chandelier at Clarendon? That might be why they stopped holding balls there,' Leonie mused.

I tried to steer the conversation back to Virginia's childhood.

'Yes . . . no, we jogged along quite nicely . . . everything seemed to be . . .' Her high-pitched voice drifted away and she left the sentence unfinished.

Leonie's distracted air encouraged my imagination to take flight. It wasn't hard to imagine Leonie leaving her child on a bus (or on the side of a road as a slightly scatty friend of mine once did in England). I could picture her wandering through the draughty rooms of that enormous house and being startled at the sight of a baby. *Oh, did I have three?* It was an exaggeration of course.

Rick was happy to admit that he wasn't a natural farmer.

'I didn't exactly fall into it. It's a tough life being a farmer. It ruins your health and your body.' As well as a triple heart bypass he had an impressive tally of broken bones and a crook back.

Was it the right thing to step back from the farm and hand over to the next generation, I wondered?

'Oh yes.'

Knowing how much conflict there must have been behind that decision I wasn't game to dig any deeper. I wondered what he might have done if he hadn't been a farmer?

'I liked jewellery and boat building. I might have done something like that . . .' He laughed. 'And gone broke pretty quickly too, I suppose.'

Art and antique sales, plus flying a light aircraft, occupied much of his time now but there was a wistful quality to his voice when he talked about staying involved in the farm.

'I still like to go out and chop wood, it keeps me active and I give some to friends in town who would otherwise be forced to buy it.'

I felt an unexpected surge of sympathy for Rick. Vee was sure of her opinions and confident enough to express them. She didn't agree with her father on many points about farming (or driving, dogs, pivot irrigators, bank loans or housing investments) and she didn't hesitate to let him know. It was like water off a duck's back. Rick remained genial and good-humoured throughout and Georgie was a loveable focal point in the midst of it all, drawing everyone's attention away from any possible conflict.

When Vee was out of the room (I wish she'd been there to hear it), Rick revealed how proud he was of his daughter's achievements.

'I took a peek at her notebook once. She had notes on some of my old ewes, suggesting ways to resurrect them by shutting them up in paddocks, it was quite extraordinary for a six-year-old.'

Rick may not have understood his daughter's sensitivities but he certainly appreciated her talent.

After lunch Vee and I drove the short distance to the weatherboard cottage she had lived in for several years when she first came back from college and was working to pay off the enormous bank loan that funded her first centre-pivot irrigator. It had an abandoned feel to it now that it was no longer lived in, paperwork strewn across an office that looked like it had been ransacked.

Vee was searching for photographs in a bottom drawer.

'That was my first boyfriend,' she said, pulling out a photo of a man on horseback. 'I met him when I worked in America.'

Vee was convinced the stint in America made her the farmer she was today. 'If I hadn't gone to America, I would have been a failed farmer.'

It was while she was still at school that Vee heard about opportunities in the Midwest. 'I knew I wanted to farm and it sounded amazing.'

She leapt at the chance to get out of Tasmania and spent twelve months living and working on a stud cattle ranch with a deeply religious family who taught her to accept the cycle of life and death.

'I couldn't get the whole Adam and Eve approach and we had hot debates around Darwin's theory of evolution, but I loved the sense of respect they had for animals, in life and in death. They taught me that for every life that ends another is born and you can't mourn every death.'

It was a formative year on many levels and the lessons she learnt went far deeper than simple farming techniques. She was taught the value of hard work, relinquished vegetarianism, suffered the pangs of unrequited first love and had a sobering reminder of how small her homeland was in global terms. When Vee's host family asked her to point out where she lived on a map of the world she pointed to Australia then stopped.

'It's not there,' she said.

'What's not?'

'Tasmania!'

The island of half a million people that Vee called home was missing. She picked up a pen.

'There,' she said, and drew it on.

She had often wondered since if she should write and

apologise to her American hosts. 'I was so hot-headed. I didn't want to do any of the boring work, I wanted to get stuck into the good stuff, straightaway!'

I couldn't help noticing that the only photographs Vee had bothered putting into albums seemed to be of dogs, further evidence (not that any was needed) of her strong connection with animals.

What Vee called her 'little hobby' when referring to dogs was in fact an all-consuming passion. Married with two children and running a farm, she still found time to train her treasured kelpies. Several were chained up outside, a reminder that these were working dogs. Her relationship with them was one of mutual trust and respect, rather than the straightforward adoration between a typical domestic dog and its owner.

'Dad's dogs are his friends whereas mine are my professional colleagues. We respect each other but we don't go and have a beer at the pub after work.'

That respect was rewarded in the 2013 Yard Dog Championships held in Queensland.

*

Vee glanced in the rear-view mirror at eight-week-old Georgie, whose tiny fist was pressed into her open mouth.

'Are you hungry Georgie? Can you hang on a bit longer?'

It was the first week of June 2013, a bright winter's day in Queensland, and Vee was on her way to the finals of the Australian Yard Dog Championships at Goondiwindi.

Earlier that day Vee had flown from Launceston to Brisbane, with Georgie on her lap and her black-and-tan kelpie, Ben, in a crate in the plane's cargo hold. She picked up a hire car in Brisbane, put the dog on the back seat with the baby and drove inland, heading 350 kilometres southwest. Ben was her oldest lead dog, one she had bred and trained since he was a puppy, and from an early age he had shown a remarkable natural ability with sheep. Like a gifted child he'd also proved rebellious at first, resisting attempts to train him. Vee had waited, patiently channelling his energy until she sensed he was ready to listen and learn. Once she'd got him onboard they became an unbreakable team; there was nothing he wouldn't try and do for her.

Her first yard dog event had taken place just three years previously, a fun day out at a farm called Williamwood where she and Ben demonstrated their obvious talent and performed well together. Vee had struggled to remember all the rules but Ben did everything she asked of him. And more. They were working in a small yard, rounding up the mob of sheep they'd been allocated, when Ben spotted another mob waiting in the wings. The eager kelpie jumped out of the pen, raced behind the shadecloth to gather up the extras then tried to round them up as well. It was a steep learning curve for them both.

Her first real competition happened not long after, at the Launceston Show, where again they performed well. At the Campbell Town Show the following year her dog Jimmy (Ben's grandson) took out the novice class and Ben won the open category. The win gave her automatic entry to the national trials the following year. Although she was a relative newcomer in the male-dominated sport, Vee knew she stood a good chance of winning.

Georgie grizzled and Vee checked the rear-view mirror again.

'Not far now, almost there.'

Highly experienced competitors were coming from across Australia to take part in three days of intense trials and Vee would be up against ninety of the best dogs and their handlers in the open category. She and Ben would get one run. If they scored well enough, they'd get a second run and their combined scores would be totalled. The handler with the highest score would be declared the outright winner. No woman had ever won the title, and certainly no woman from Tasmania.

She'd left Henry with Steve back in the Meander Valley, where there were still a few weeks until the main calving season hit. His parents were doting grandparents, always willing to help, and her parents were only an hour away so Steve would have plenty of help if he needed it. Georgie was breastfeeding so Vee brought her along. It was typical of her have-a-go approach that she decided not to let a newborn baby stand in the way of a tilt at the national title.

She checked into the local caravan park, fed Georgie, let Ben out for a run and met some of her fellow competitors. As she suspected, most were men. Thankfully there were plenty of wives and girlfriends on hand as well, and the women clustered around the baby.

'Would anyone mind holding Georgie while I do my run tomorrow?' she asked. A raft of hands shot into the air.

Vee cast Ben, sending him out into a small yard to gather a mob of sheep and manoeuvre them through a series of gates and around obstacles. She used simple, short commands.

'Walk up! Over! Sit!'

Their first run was near-perfect, with no time penalties and a very high score. She waited to see how the other dogs would go, quietly confident they'd done enough to make it through to the final. The judges conferred, scores were tallied and Vee and Ben were through to the last seven.

'Next up in the final round we have Vee Chilcott from Tasmania, with eight-year-old Quamby Ben.'

Vee stood up and led Ben confidently into the yard.

'Sit!'

They waited for the bell signalling the start of the final run.

The result would hinge on what happened in the next ten minutes.

'Walk up!'

Ben got to work, performing each task with pinpoint accuracy, starting the sheep then stopping them, losing no ground as he followed Vee's every command. Once again, it was a perfect round. The minutes ticked past until all that remained was one last gate.

'Sit!'

Ben dropped to his haunches and Vee walked up to the waiting dog. The sheep gathered in front of them were perfectly positioned for the final gate. All she had to do was 'send' Ben and the title was theirs. Then, for some inexplicable reason, she blew it. Instead of sending Ben in the direction he had been going she ordered Ben to come back around her. The surprised dog did as he was told, the sheep moved onto the gate and Vee had to correct him, sending him around again to move the sheep back into position. In that single manoeuvre the vital points that would have seen her declared the first Tasmanian – and the first woman – to win a national championship were lost.

Ben knew instinctively something had gone wrong. His ears flicked up and he looked at Vee with wide eyes, as if to say, 'I know something went wrong, but what? What did I do? What happened?'

Vee was bitterly disappointed in herself. She knew the fault was hers. At the last minute she had changed her normal routine, thinking that was the way the judges would prefer to see the manoeuvre done. As a result she lost critical points, and the title.

It's no exaggeration to say that Vee felt she let her destiny down that day.

'It was one of those "if only" moments. The stars were all aligned in Queensland and I blew it. One day I might forgive myself.'

It was an achievement to be placed fourth out of a field of ninety competitors in her first national yard dog trial (the event was won by defending champion and veteran handler Joe Spicer with Gogetta Package) but the genes that saw Vee's grandmother develop a thriving business in a highly competitive man's world won't let Vee rest on such flimsy laurels.

There would be other opportunities and other competitions, and I had no doubt Virginia Chilcott wouldn't make the same mistake twice.

*

As we drove back towards Meander I asked Vee about her experiences in the immediate aftermath of taking over the farm. It was what she'd been longing to do. Left in sole charge of an 1100-acre property she went all out to make her ideas work, struggling in the first few years as she invested everything she had (and plenty she didn't) in costly improvements that she hoped would pay dividends in future years.

More than anything she was keen to improve the water infra-structure. (Cue those pesky soft hoses she hated so much.) If her idea of planting a commercial crop of wheat, and possibly other commodities, was to stand any chance of success she needed a reliable source of water, and that meant replacing the soft-hose systems her father and grandmother had used.

'What's wrong with using soft hoses?' her father asked.

'They're labour-intensive and uneven.'

'We've never had a problem with them.'

'Really? How come your poppy crop failed?'

'Who says it failed?'

'You know it did! Look, I need a reliable source of water and I'm not going to waste my energy on stupid pipes that blow out and ruin the crop.'

'Why do you want to plant wheat anyway?'

'So I can sell it.'

'I thought you wanted to farm sheep.'

'I do! I can do both if I have a centre-pivot irrigator.'

'It's a lot of money to spend.'

'I know and it's the only way forward if I want to improve productivity.'

The arguments continued for weeks until Vee finally overcame her father's objections and he agreed to step in as guarantor for the $156,000 she needed for the centre-pivot irrigator. Rick may have bowed out of the business but his purse still had money in it, and that gave him some measure of control.

Of course, once the centre-pivot irrigator had proved its worth any objections were forgotten.

'I'm glad we got that irrigator, it works well, doesn't it?' her father said one day, which only served to irritate Vee further.

'It was my idea, not yours remember!'

Several years later, as a mum with two young children, life had changed dramatically.

Henry was born in March 2011 after a difficult five-day labour followed by an emergency caesarean as his heart rate dropped. 'It was the most unenjoyable time of my life. There was nothing romantic about the birthing process.'

Maintaining her productivity after Henry arrived proved impossible, not that she didn't try. Within weeks of giving birth, Vee was back at Epping Forest, working up the ground to plant wheat. Henry spent his first month bouncing around in 'a rough old tractor nicknamed The Fossil'. Much to her dismay, breast-feeding only lasted six weeks. 'It was a combination of bad management and not knowing what I was doing.'

Vee struggled to cope. She was still paying off the first pivot irrigator and, in order to maintain her productivity, she put Henry in day care five days a week. 'He got fractious, I got bad-tempered and it all got too much.'

Reluctantly Vee was forced to accept that she wouldn't be able to work the way she used to. Henry's arrival hit Vee hard financially too. With two businesses the couple had kept their finances separate, which meant mince instead of steak whenever Vee shopped for groceries. 'My pride wouldn't let me show Steve how much I was hurting. I simply couldn't afford to buy steak.'

Two years later Georgie was born (an elective caesar this time) and Vee was back at work preparing the ground for another crop of wheat just days later. With finances still tight, there was no money to pay anyone else to do the work.

'The farm at Epping Forest is only 1100 acres so I never had enough work to justify paying someone.'

Like her brother before her, Georgie was strapped into the tractor in a car seat while her mum worked, but at least it was a new tractor this time, a John Deere 6930 Premium. Encouraged by Steve, Vee had ditched 'The Fossil' and taken out a loan for a new one. She'd also invested in a second centre-pivot irrigator. It meant having to find over $2500 in repayments each month just to satisfy the tractor loan, so there was still no fat in the budget.

'Steve subsidised a few repayments until I had lambs ready to sell.'

Her Epping Forest property may have looked extremely productive and prosperous but obviously looks could be deceiving. Vee had inherited a family business and, like so many farmers, she had been forced to make huge investments in order to modernise it and stay ahead of the game.

We picked up Henry from day care, dropped into the supermarket in Deloraine to pick up groceries and drove on to Meander. Having met her parents and heard a little about her childhood, I wondered if Vee would do anything differently when it came to raising her own children.

'Yes, I think so. I hope I can be more explanatory, not on little things but on the big-ticket items.'

In spite of the obvious differences between sheep and dairy, Vee and Steve have found a surprising amount of synergy between their two farms. Over dinner that night we talked about how in the past couple of years some of Steve's cows had overwintered at Epping Forest, avoiding the wet mud of Meander to enjoy a six-week break on sunny paddocks full of rich grass. And thanks to Vee, Steve had discovered how useful a well-trained working dog could be.

'I gave him one of my old work dogs and she was smart enough to overrule Steve if he gave her the wrong command.' For the first time, Steve no longer needed another employee to help him draft pregnant cows or move frisky calves.

Vee had started a liquid fertiliser spreading business (the spreader originally financed by Steve and since paid back), which was easier to run with two young children.

'I jump into the tractor with Georgie or Henry sitting in the little seat beside me and it's like a breath of fresh air. I didn't know making money could be so easy!' (I doubted if it was as easy as she made out.) Not surprisingly one of her main clients was Steve's Meander dairy, which got charged 'mate's rates'. And it was Vee who encouraged Steve to use sex-specific semen to cut down on the number of unwanted male calves.

'I think the bobby calf industry is cruel. Bull calves are surplus to requirements so they get turned into veal at only a week old.'

Sex-specific semen (most of it from France) ensured a higher percentage of female calves. Vee's concern for animal welfare and a lingering reluctance to kill lay behind the push.

Steve's most recent innovation was to move the dairy from a conventional operation to a biodynamic one. 'We used to push grass to the maximum, putting on urea every five weeks, feeding the grass to feed the cows to produce more milk . . . yadda, yadda. It was unsustainable.'

I probed further and he admitted to a twinge of fear that such a high chemical input might have been poisoning the very animals he was trying to look after.

'We'd have sick cows for no reason, maybe ten during the year, they would just go off their milk and four or five would die and no one knew why.'

Following the advice of Australian dairy consultant Dan Huggins and American guru Hugh Lovel, Steve changed to a biodynamic operation and the dairy reaped the benefits almost immediately.

'Within three months we could see a difference.' The fertiliser bill had since been cut in half and sick cows were a thing of the past. 'The more we correct the imbalance of nutrients in the soil, the less fertiliser we have to use.'

An annual crop of grass, maize or rye corn was sewn in November, harvested in April/May and turned into silage to fatten the cows, the land now so fertile he could reap a large tonnage off a small acreage.

'We're looked on as a bit weird, I suppose, but others have seen what we've done and they're starting to realise there might be a market for milk that is produced biodynamically.'

Steve's father used to be flat out milking 180 cows, and if water supplies dried up around Christmas time it was a battle to produce any milk at all. The creation of Meander Dam in recent years helped ensure a consistent flow of water through the valley but it was the changes Steve introduced when he took over the dairy at age nineteen that had the greatest impact. He invested half a million dollars on pivot irrigators, he completely revamped the dairy platform and he doubled the size of the herd, shifting the entire operation up a gear. His father, who was still very involved in the business, was happy to let his son take the lead.

'It was all I ever wanted to do, I had an affinity with it,' said Steve.

Just like Vee, he couldn't imagine doing anything else but farming.

Snow was forecast the next day and wood smoke curled from chimneys in the lush green valley as we finished breakfast. Henry had already been outside for almost an hour that frosty morning to help feed the peacocks, check on the dogs and chase the cats, scooting along the veranda on his tricycle. 'It won't be long before he tackles that hill,' Vee said, watching her fearless son race across frozen puddles, searching for one with enough water to make a decent splash.

The twisting road that led down the valley would be an obvious attraction for Henry and his tricycle, although if he took after his father he would sooner be heading up the hill than down it. Steve spent his youth hiking, shooting, fishing and camping high up in the Great Western Tiers behind the dairy. His father would drop him and his fifteen-year-old friend Jamie off with their tents and cooking gear then pick them up at a pre-arranged spot several days later. Their motto was, it's only trespassing if you're caught.

Would they allow Henry the same freedom?

'I don't see why not,' said Steve, as Vee bundled the protesting Henry into the car. 'Things have changed a lot nowadays, but not up in the mountains.'

A quick detour up to see Meander Dam before Vee took Henry to day care revealed a pristine wilderness and a vast body of water.

'I know some people get upset at the flooding of a dryland

forest but this is far better for farmers because it controls the environmental flows.' I didn't know enough about it to comment. I just knew I was in an area of outstanding natural beauty.

We drove on to Deloraine,where Henry attended day care three days a week. Sometimes Vee took him at seven-thirty, sometimes not until ten-thirty, it all depended on what needed to be done; their farm work was regulated by tasks, not hours.

From Deloraine it was another seventy-five kilometres to Epping Forest, and it was mid-morning by the time we turned onto the six-kilometre road that led down to Glasslough.

Vee pointed out a modest house at the top of Bellevue Road that I hadn't spotted the day before. It was where her grand-mother and grandfather lived after they vacated the main property to make way for Rick and Leonie.

There had been talk of Vee's parents doing the same when Vee took over the farm, then she met Steve and it became clear that their main residence would have to be at Meander. Talk of any handover of the family home had been quietly shelved for now.

In front of the small house where Vee's grandmother lived was a dry stone wall stretching for fifty metres or more.

'My grandmother built that by hand when she was in her late eighties. She shifted 300 tonnes of rock,' said Vee, with under-standable pride. It wasn't hard to imagine Vee doing the same.

Early July was normally when Vee would be lambing, worrying about wet paddocks, pre-vaccinating and staying up late to finish the job no matter what the time might be, but that was in the pre-Steve, pre-baby days.

'Now that we've got Georgie as well as Henry I've given myself permission to put less pressure on myself. It's the first time in my adult life I haven't had to worry about lambing.'

We stood in the centre of a paddock on a north-facing hill, surrounded by contented cows grazing on rich grass in the winter sun. Most had been sent down from Steve's dairy and others were on agistment from other farms in the Meander Valley. Some of the cows were so full they'd stopped feeding to lie down and bask in the warm sunshine. 'They're all looking good,' she said.

Vee waited for Georgie to finish breastfeeding then reeled in the electric fence to move the herd of pregnant cows onto a fresh paddock. 'That's Jenny Craig,' she said, pointing to the strip they had just come from, with hardly any cover. 'And this . . . this is McDonald's!'

She wound up the live wire and the cows clambered to their feet. 'COME ON, COME ON!'

They bellowed softly in answer to her call then lumbered forward as the line of electric fence disappeared and the pregnant cows realised there was fresh grass to be had.

Steve wintered over half his herd at Epping Forest that year and the following year he was planning to send more. The paddocks at Vee's farm had plenty of cover to fatten the cows before they were taken back to the colder, wetter microclimate at Meander, where most of them would give birth.

Again I was struck by the sheer size of the docile beasts in front of us.

'These cows are at the limit of what they can produce but consumers want cheaper milk and farmers have to make money somehow,' Vee said. 'They've had to breed cows to produce as much milk as they can.' So I'd been right about the size then. These cows *were* bigger than I remembered.

I wondered how farming might change for her children's generation.

'I don't know. Will we keep trying to increase production? Keep intensifying what we do? When you see a cow's udder in full milk production it looks unnatural, but the world has said,

"I only want to pay a dollar for a litre of milk." Last year farmers were paid thirty-five cents a litre for milk that cost them thirty cents a litre to produce.'

Those margins sounded impossibly small; some reports I read later suggested dairy farmers were going backwards, with the cost of production outstripping income. It was a dispiriting thought.

Like her grandmother before her, Vee was passionate not just about animal welfare but about the business of farming. She was closely involved in industry matters, having been a long-standing board member on the Tasmanian Farmers and Graziers Association and a director on the Sheep Council of Australia.

With the arrival of two children, Vee had been forced to change the operation of her farm to make it easier to manage.

'There was no point dragging two businesses down,' she said, pragmatically. Her farm had returned to Merino wethers, producing wool, much like her father and grandmother 's farms before her. It was a less labour-intensive operation than prime lambs, and lighter stocking rates meant she didn't have to feed in winter. She'd kept 100 or so crossbred sheep to supply lambs for a fledgling business delivering prepared lamb directly to households.

'It's great, perfect for while the children are small. People like to know where their meat comes from and I can take the children with me on delivery runs.'

When Georgie and Henry were older she would be able to give the farm more attention and crank up production. For now, though, the liquid fertiliser business, overwintering cows, running Merino sheep and some cropping kept her busy enough.

Both she and Steve struck me as highly motivated farmers as well as astute business people. Yes, they'd had the good fortune to inherit part of their respective family businesses but neither of them took that for granted. They recognised that investment would always be needed if those farms were to thrive and prosper for the next generation.

It would be a few years yet before either of their children was faced with the choice of working on the land or leaving it.

'As long as they have a career choice, I don't mind what they do.'

With a combination of Prevost and Chilcott genes running through them, it was likely to be farming.

Lyn and Rob French

'Gilberton'
Approximately 450 kilometres inland from Townsville, Queensland

A single phone call to Lyn French confirmed what Kate Philipson, communications co-ordinator for the Royal Flying Doctor Service in Queensland, had told me. 'She's a great storyteller,' Kate said. And she was.

Lyn was a gift to any writer – open, friendly and willing to explore some of her most personal demons. She had great stories to tell and she wasn't looking to sugarcoat any of them.

'Where exactly do you live?' I asked, having failed to find Gilberton Station online; all Google maps could show me was a town over a thousand kilometres in the wrong direction.

'Don't worry,' said Lyn. 'I'll send you a mud map.'

The sun rose on a clear winter's day as I drove inland from Townsville, a crumpled mud map lying on the passenger seat beside me; I wasn't even convinced it was the right way up. Mobile coverage stopped about sixty kilometres inland and sat nav gave up shortly afterwards. 'Unknown road,' was all it said, discouragingly.

Semi-trailers and tourists towing caravans occasionally hove into view as the road cut through the Harvey Range. Two hours later there was a roadhouse, two hours later another, then all traffic disappeared. For the next three hours the isolated dirt road bucked and twisted, swooping into dry creek beds and soaring up the other side, wheels slipping on grit, sand, stones and dirt that switched from chalk-white to the colour of dried blood, then back again in an instant. I didn't see a single other vehicle.

The effects of long-term drought were everywhere; in places the winter sun blazed down on country that had been flogged and overgrazed, leaving nothing but rocks and bare earth with forlorn cattle searching for feed. Elsewhere the dry brittle grass looked rich by comparison, but it was still dry and brittle. Termite mounds as tall as kangaroos lurked at the side of the track; sometimes those mounds *were* kangaroos, exploding into view with a single bound. Black feral pigs burst out of the bushes, the direction their nuggety energy would take them impossible to predict. Eagles fought over roadkill, tearing at the glistening flesh until my approaching vehicle forced them airborne with a muscular flap of heavy wings. There were emus, rabbits and rock wallabies, and the longer I drove the more ubiquitous were the solemn faces, long ears and loose skin of grazing Brahman cattle.

Towards the end of the journey I started to doubt the mud map Lyn had provided. I'd forgotten to zero the mileage at the most recent waypoint and it had been three hours since the last sighting of any other vehicle. Then the road crested a steep rise and the sun glinted on the window of a tiny caravan in the distance, parked at the bottom of a sandy creek bed. It was an incongruous sight: an old-fashioned 1950s caravan with rounded edges, barely large enough for a single person to stand upright in. An old fella in dirty shorts was hunkered down beside it, staring intently at a patch of dirt.

He looked up as I got out of the car and waved to him.

'Found any gold?' he shouted across the intervening gully.

'I'm looking for Lyn French,' I called. 'Am I on the right road?'

He nodded and waved me on. 'See any gold, you let me know,' he shouted, and bent back to his task.

Twenty minutes later I passed a handmade sign, hammered onto a post that had been driven into the rocky hillside.

'Speed limit 15. You hit my kid or dog because you're speeding you won't need a lawyer.'

Can't is a dirty word

My first thought when I saw that battered sign was to check my speed. The implied threat had me worried. What kind of family was I visiting? (It was only later that I wondered how many 'passing motorists' they would ever get in such a remote spot, and then, how many of them would be likely to be speeding? As it turned out the answer was plenty; gold prospectors are an eager lot.)

When I phoned to set up the interview I had asked Lyn if she thought I might need a satellite phone.

'Nah, you'll be right,' she'd said.

Seeing how far inland I was going I played safe and splashed out on an expensive satellite phone rental, exposing me as a nervous city slicker equipped with a satellite phone, space blanket and head torch in case of emergencies. On the passenger seat, as well as the crumpled mud map, I had supplies of bread, water, bananas, chips and cheese – a city girl's idea of emergency survival items – plus wine as a gift for my host (which I'd already decided I would drink if I ended up in a ditch). Dire warnings from people accustomed to dirt driving had convinced me that was my likely fate. *Always break before a bend; if you break on the bend you'll roll the car. Never swerve to avoid a kangaroo; hit it head-on or you'll roll the car.*

The prospect of rolling the car stayed with me most of the journey. After seven hours my fingers were tingling from their tight grip on the steering wheel. It had been a challenging drive and, far from being alone, I'd had plenty of company, just none of it human.

My visit to Adele and Philip Hughes had taught me some rudimentary knowledge of cattle but I was under no illusions about my ignorance. I recalled telling Philip I would be visiting Lyn French and he knew instantly where she lived.

'Up past the Oasis Roadhouse, in behind Einasleigh. We had a block up there once, it's pretty rough country.'

'Yes, apparently all she can do is fatten,' I said.

Philip let out a big laugh. 'I think she'd be pushing shit uphill to try and do that,' he said. 'I think she would have said all she can do is breed.'

'Ah yes, that was it,' I said, chastened yet again by my woeful ignorance.

The aggressive warning about speeding unnerved me so I slowed down even further. Rounding a bend I came across the first gate I'd seen for hours. Cattle ambled out of the way as the car approached, kicking up dust to add to the trail left behind. Beyond the gate the track rose steeply, past a breezeblock dwelling and up a tree-studded hill. Towards the top, two dogs appeared, barking furiously as they kept pace with the car, and finally, seven and a half hours after leaving Townsville, there was Gilberton Station – a single-storey timber-clad building painted buttercup yellow, perched on top of the hill.

'You found us all right then?'

Lyn French's high-pitched laugh would have shocked a kookaburra.

Sitting on Lyn's wide veranda later that day, nursing a welcome cup of tea, I took in the view. The land around Gilberton is rocky high country with thin clean air – over 500 metres above sea level – and it was a typical winter's day: sunny, cold and dry. On all sides, in every direction, stretches of tree-topped hills sat quietly under a blue sky devoid of cloud. Night-time temperatures in winter can drop as low as minus three and summer is usually hot and wet, extreme tropical downpours turning the Gilbert River into a raging torrent. I'd driven across that raging torrent earlier; it was a dry sandy bed, with no water in sight.

Having learnt my lesson from the previous visit to Queensland and I was rugged up in jeans, t-shirt, sweatshirt and scarf. Lyn was barefoot, her only concession to the cold a sleeveless gilet worn on top of jeans and a pale blue shirt. She was short, barely five foot two, and I had no doubt that her small frame contained more energy and power than most women twice her size. We sat chatting on a sheltered veranda, surrounded by handmade wooden furniture and battered cooking pots that hung from a long slab of timber as Lyn recounted her extraordinary story in blunt, unemotional language, punctuated by that wonderful high-pitched kookaburra laugh.

Lyn French was born in 1966 in the Lockyer Valley of South East Queensland, where her father ran a produce agency. While she was still a toddler her father gave up the agency and bought into a furniture removal business on the Sunshine Coast, only to move again a few years later when he bought a cattle station near Rockhampton.

'We thought it was exciting to be moving to a property. It was a big place because my father always wanted bigger and better than anyone else.'

Kindly neighbours taught them a lot about cattle and by the time Lyn was twelve her father had shifted the family again, this time to Gorge Creek, 500 kilometres inland from Townsville, where her father bought 150 square miles of tableland country just behind Gilberton, not too far from where we were sitting now.

'We thought we were cattle producers by then, we thought we knew it all. Compared to the land near Rockhampton, it was really just a big heap of shit.'

The oldest girl in a family of six children, Lyn recounted in a matter-of-fact way how she and her siblings were forced to work from an early age. 'The old man was pretty tough. I could work cattle without any fear but I never liked going mustering. I was terrified of horses. He'd chuck us on a horse then slap it and if you fell off you got a flogging.'

Physical punishment was handed out by the strict Christian for every misdemeanour, from answering back to failing to learn to drive. Lyn fought against her strict upbringing and constantly clashed with her parents, especially her father.

'If they said it was white, I said it was black; if they said green, I said pink. When we were young it got to the point where Mum had to dress us all in long pants so no one could see the bruises.'

To the outside world, though, they were a model family. 'We went to church every week and I used to think, if there really is a God, he sure hasn't done a good job of looking after us.'

Lyn took refuge in machinery. 'I learnt to drive young and pretty quick when the alternative was a flogging.' She spent most of her waking hours on one of two TD14 'dozers' – old diesel-engine tractors, 'like prehistoric monsters', that were used to push over trees and pull ploughs to plant sorghum. When a neighbour bought a new dozer, Lyn's father went out and bought a bigger, better version – an FD30, about ten foot high and twenty foot long. Lyn drove it for hundreds of hours, day and night. 'I thought I was pretty cool.' I wondered if she could even have seen over the steering wheel.

School proved no escape. Lyn remembered starting grade one, she could even recall her first teacher's name – Mrs Pickersgill – but by the time she got to grade seven she had only completed a total of four years' tuition. 'We were always being pulled out of school to work.' Living where they did, the only option for high school (other than boarding school) was by correspondence. Fate intervened to put a stop to that too.

'My mother had a bad horse accident that landed her in hospital for three months and it all went out the window then. We just didn't do school anymore, not that I can remember.'

I sat quietly listening to all Lyn had suffered, something I had no experience of. At one point she said, 'There was no love in my childhood.' What an awful picture those blunt words conjured.

At fourteen, Lyn ran away from home, determined not to let her life be ruled by her strict father. 'I was sick of the floggings and life seemed . . . I just thought, what's the sense? I reckon Mum was looking for a way out as well but I wasn't going to stick around to find out.' She left vowing never to get married and never to have children. 'I couldn't see the sense in having kids if all childhoods were going to be like mine.'

Lyn left in the company of cousins who'd been staying on her parents' remote cattle property. She persuaded them to take her back to Charters Towers with them. From there she jumped a train (with no money and no ticket) to Barcaldine, where an uncle found her a job in the shearing sheds, picking up fleeces.

'No one tried to persuade me to go home.'

I wondered if she wished they had.

'None of the stations had phones anyway, so there was no way of getting in touch with me and, besides, I didn't want to be found. I had my life to live.'

By fifteen Lyn was working full-time as a roustabout. 'It was shit money but I felt like I was a millionaire. I was unbreakable!'

For a while Lyn lived on a station owned by an elderly woman, Agnes Pumpa, helping in the garden and doing general maintenance around the house. 'She would give me a list and send me to the shops and I'd just guess at what she wanted.' If any items were missing off the shopping list, Lyn would lie and say they didn't have it in stock. 'I was a great bullshit artist. If she asked me to read anything I'd say I've got something in my eye, I can't make it out.'

So when Lyn ran away from home at fourteen she hadn't been able to read or write. I wanted to know more. She sounded confident and articulate and the only indication of language issues was an impressive line in swearing. How had she coped without being able to read or write? Before I could ask any questions she changed the subject. I made a mental note to go back to the issue later.

'Me and Julia, we were soul mates, inseparable.'

'Julia?'

'Julia Whela. She was adopted. All her life she'd been trying to find her real mum because her adoptive mum had died. We met in Barcaldine and we became great friends. Every weekend we'd go travelling around the rodeos and if somebody bet us fifty bucks to do something stupid, we'd do it.

'There was a B&S [Bachelor and Spinster] ball coming up in Barcaldine and some guys bet me I wouldn't climb the windmill in my ball gown.'

Lyn may have been a daredevil but she was canny with it. The night before the ball she climbed the windmill while she was sober, just to make sure she could do it, and the next day she won $250 for repeating the feat, drunk, holding a glass of rum without spilling it, wearing a ball gown and high heels.

'I didn't care that it didn't look pretty, I got my money.'

She lived from pay week to pay week, blowing money up the wall as the mood took her, with her ever-present confidante and soul mate, Julia.

'We were made of steel. We both said we'd never get married and we vowed we'd never have kids.'

In between drunken pranks, the girls grew closer. Lyn shared secrets with Julia she'd told no one else, and Julia involved Lyn in the search for her birth mother. When Julia finally tracked her mother to an address in Sydney the girls arranged to fly together for what promised to be an emotional reunion. The meeting never took place. Just weeks before the trip, Lyn's best friend was killed in a horse riding accident.

'I went off the rails. Julia's death made me crumble. She was just such a great friend. I didn't do drugs or anything, just lots of smoking and grog. One day a friend took a video of me. I saw myself and I thought this is bullshit. I have to stop drinking.'

Lyn sobered up and got back in touch with her family.

'Dad's business had gone bankrupt. He was making a bit of money on a gold mine near Gilberton and, I don't know how, but he talked me into coming home to drive a truck on the mine.'

Which is how Lyn ended up marrying Rob French.

Rob's ancestors, the Martell family, arrived at Gilberton in 1869 as teamsters, bringing goods to supply the largely Chinese population of prospectors who had flocked to the area in the Gilbert River gold rush. Almost immediately the Martells started a butcher's shop and took up 88,000 acres at Gilberton – a small property compared to most in the area – to breed cattle to supply the shop.

Lyn had 'noticed' Rob French when she was just thirteen and still living at home on Gorge Creek, the property behind Gilberton Station. Rob was fifteen at the time and a fifth-generation French who, like his ancestors, lived and worked on the land. Rob joined team musters at Gorge Creek on a couple of occasions and Lyn remembered being impressed enough to mount a horse and go mustering with him, just to check him out.

Her overall impression was of a big shy bloke. 'He was a timid mouse who would only ever look at me from under the brim of his hat, out of the corner of his eye, then he'd give me this cute smile.'

Their mutual interest didn't go unnoticed, prompting Lyn's father to threaten the shy boy with a brutal warning, 'Hey, young fella! You keep your eyes off my daughter or I will cut you, you hear?'

Rob was nineteen and Lyn eighteen when she went back to live with her family in 1985, by which time she had vowed never to get married and never to have children. Even if she hadn't taken such a drastic vow, Rob could never have been a suitable candidate for marriage. 'He wasn't in our religion and my father was very strict about things like that.'

Which made it all the more surprising when Rob arrived at the North Queensland gold mine where she had taken a job driving a truck.

'He just turned up.'

The next day Rob turned up again, then her shy next-door neighbour started talking to her.

'He came out in the truck with me every day after he'd finished work. Sometimes he'd stay all night then he'd go home to work the next day.'

Three weeks later Rob proposed.

'What did you do? What did you say?'

Lyn laughed at my eager questions (I've always been fascinated by how people get together). 'I said woah, boy! Why don't we go out for dinner first?' What a sensible woman.

A romantic dinner wasn't so easy when the nearest restaurant was 480 kilometres away but Rob made the arrangements and Lyn sensed there was something special about him. Although he was painfully shy, Rob had a deep well of inner self-confidence, having grown up in a loving family environment with parents who nurtured and respected him. His was a very different background from Lyn's.

'But how did you know he was the one?'

'I just knew. In spite of the vow I'd taken, I just knew.'

A high-pitched giggle suddenly exploded out of her tiny frame.

'Then I made him wait nine months because all the old biddies around here would have thought the only reason somebody would get married quickly was because they were pregnant! It was bedlam for him, it really was. But we did it!'

In the months that followed the engagement, Lyn's parents encouraged Rob to attend church, hoping he would join the religion. He dutifully went along to keep the peace until one day he turned to Lyn in church, his voice calm and low.

'I will never stop you going to church if that's what you want to do, but don't expect me to go once we get married.'

Lyn was delighted.

'It will be the happiest day of my life when I can walk out of this place and never have to go back!' she whispered.

Neither of them was looking forward to a church wedding, Rob because he hated the spotlight and wanted to elope and Lyn because her parents' religion forbade alcohol, make-up and loud music.

'The old man was paying so I thought great, this is going to be boring!'

In the end they reached a compromise – a garden wedding at Tinaroo Dam on the Atherton Tablelands, with a singer playing acoustic guitar. When the wedding was over Lyn and Rob set up home together on Gilberton Station, in a caravan parked high above the Gilbert River.

As the temperature dropped we moved the conversation inside, although not before Lyn had taken me on a tour of her drought-tolerant garden that stepped up the hillside in a sweep of flourishing cactus interspersed with rocks and rusting machinery. At the top of the garden was a covered lookout, furnished with rustic chairs crafted from twisted timber branches – a perfect place for an evening beer.

'If plants grow for me they're tough. I like plants that don't need water or maintenance, they don't get any TLC here.'

It was a far cry from the high-maintenance requirements of my small inner-city courtyard and a sobering reminder of how precious water was.

'Where does your drinking water come from?'

'The river.'

Unseen in the deep gully below was the Gilbert River, or at least the dry bed where the Gilbert River would hopefully be running in a few months time. Living in a city, I rarely thought about water supplies unless there were drought restrictions. On the Rob and Lyn's property all the water for cattle and human consumption came from dams or from the Gilbert River. When surface water dried up, as it had now, they put down a sand spear, like a pipe with fine mesh on it.

'We don't treat it and we're all healthy.'

It was too late to wonder if my city-based, chemical-intolerant vegetarian constitution would survive drinking river water; I'd been drinking it since I arrived. Que sera, sera, I thought.

Lyn's husband was in Townsville the night I stayed, picking up a load of steel and iron on two semi-trailers, and he wasn't expected back until the following evening, so we continued our conversation after dinner, ensconced on comfortable sofas in the open-plan living/dining room that incorporated a kitchen with a wood-burning stove – the only form of heating in the house. Kangaroo skins hung on one wall, handmade quilts lay on the back the sofas, and a wealth of framed family photographs suggested a close, loving relationship between the generations.

It wasn't quite such a rosy picture at the start of their married life together.

In 1986, Rob's parents were living on Mount Hogan and Lyn and Rob were on the adjoining property, Gilberton Station – in terms of distance the equivalent of driving from Sydney Opera House to the centre of Parramatta. That qualified as the back-yard as far as Lyn was concerned. Having done everything she could to escape her own family, it's fair to say she wasn't too keen on living near her in-laws.

'I was horrified. I wanted Rob all to myself. I didn't see why we had to have anything to do with his parents.'

Lyn's rebellious insistence that she never wanted children didn't last long. Within six months of the wedding she was preg-nant. Living a seven-hour drive from the nearest hospital meant Lyn had to leave home six weeks before her due date. She spent the final weeks of her pregnancy with her parents in Malanda, six hours north of Gilberton, and Rob got a job in the local milk factory to be by her side. Kerri-Ann was born in nearby Atherton in December 1987, after a scant hour-long labour.

Ashley was born the following year, in December 1988, by which time Lyn's parents had moved to Charters Towers. More relaxed the second time around, Lyn waited until four weeks before her due date to leave Gilberton and Rob went with her again. She vividly remembered waking him up at two o'clock in the morning.

'Rob, I think you should take me to hospital.'

'Can't you wait until morning?' he mumbled.

'No.'

Rob drove his wife to hospital and twenty minutes later he was back.

'Why aren't you staying?' Lyn's worried parents asked.

'She's had it,' he yawned. 'Little boy, she's gone to sleep.'

When it came to Anna's birth, four years later, Lyn was so relaxed she waited until she was thirty-six weeks before they left Gilberton.

'You've seen calves being born,' she said to Rob. 'You can do it if you have to.' Much to Rob's relief, their third child, Anna, was born at the Kirwan Women's Hospital in Townsville in December 1992.

With Anna's birth came grim news.

'I had cervical cancer before she was born. They thought it was fixed but it came back real bad.' Rob proved more than just a shoulder to cry on. 'He came with me to every doctor's appointment and every treatment. He was my soul mate. If I hadn't had Rob, with his gentle, caring nature, I don't know what I would have done. There's no sick pay out in the bush and this was in the drought years, when cattle prices were shit.'

I tried to imagine what lay behind those words, the untold hours of driving across dirt roads to get to the nearest hospital with radiology and chemotherapy facilities, the worry of having three young children and no income, the fear of not knowing as she held her newborn if she would survive to see any of them grow up. There was no point pretending I had any idea of what Lyn and Rob must have gone through.

Lyn's parents had split up by that stage and she kept the diagnosis secret from all but her younger sister, Bec. The only other person she leant on was her father-in-law, Gus. Thirty years previously, Rob's father had faced a similar crisis when he'd been told to go home and put his finances in order after doctors diagnosed late-stage melanoma.

'He fought it and he won because he wanted to see his sons grow up.' There was a fierce, loving energy in her words. 'He was a great strength during that time and I grew real close to him.' There was a fierce, loving energy in her words.

She shook her head. 'It's so funny, I thought my throat had been cut when I first got married, having to live so close to my in-laws, and I've learnt so much from them. It was just selfishness on my part. I'm older and wiser now, although it took me eight years before I woke up to myself and realised they were just loving parents who wanted the best for their son. I'd never seen that closeness in a family before.'

Lyn talked me through the photographs hanging on the wall. One taken in December 2012 celebrated her parents-in-law's fiftieth wedding anniversary, Lyn and Rob's twenty-fifth anniversary and their son Ashley and his wife Camilla's first wedding anniversary. The toddler in the photograph, another Robert French, was the seventh generation of the Martell/French family to be living on the land at Gilberton.

'He's a cute little man and he doesn't like wearing shoes either. It's a family trait.'

Other photographs showed previous generations, including Rob's great-great-grandmother, a second-generation French who was born on the property.

I was wondering if now might be the right time to ask about schooling, about how Lyn felt trying to teach her children when she couldn't read or write herself, but her next sentence stopped me.

'We lost a lot of photographs in the fire.'

In 1994, two years after Anna was born, Rob's parents had helped finance the building of a proper brick house (up until then, Lyn, Rob and the children had gone from living in a caravan to a basic shed built from hebel blocks, with a couple of bedrooms, a bathroom, a toilet and a small kitchen). Thanks to Rob's parents they had moved into a substantial home, with a schoolroom out the back, a large eat-in kitchen, a proper dining room and bedrooms to spare.

They lived in the house for seven years and then, in the blink of an eye, it was gone.

*

The fire started late one afternoon while Rob was putting out lick (a supplementary feed) for the cattle, helped by the two youngest children, Ashley and Anna. Kerri-Ann, who was fifteen, was helping her mum fold washing and had just been over to start the generator at the back of the house when she came in and sniffed the air.

'Can you smell something funny, Mum?'

'No, what sort of funny?'

'Like burning.'

Lyn checked the wood-burning stove, just in case, but nothing seemed amiss so they carried on with their chores. A few minutes later, Kerri-Ann walked into the bedroom with a pile of clothes and saw smoke coming from the ceiling.

'Mum! The ceiling's on fire!'

Kerri-Ann raced outside to turn the generator off and Lyn grabbed the phone. She just had time to ring Rob's father, Gus, before the line went dead.

'Call your father on the radio,' she shouted.

Lyn pointed a hose at the burning roof but within minutes the fire had turned into a blazing inferno. She and Kerri-Ann grabbed what personal items they could and threw them out of the house.

'The cabinets! Mum!'

The cabinets Kerri-Ann was referring to were antiques, inherited from Rob's great-grandparents. The cabinets themselves were valuable enough, but more importantly they were full of trinkets from generations back, as well as some of their family photographs, including the negatives.

'Quick, grab that end.'

With burning embers falling around them Lyn and Kerri-Ann lifted the first heavy cabinet and carried it outside then raced back in for the second and dragged it onto the veranda as the blaze took hold. By the time Rob got back from the paddocks, the house was reduced to a pile of ash on the ground.

Seeing the veranda bearers were still smouldering, Rob and his father went to move the cabinets. They couldn't even lift them. Adrenaline had given Lyn and Kerri-Ann the strength to save all that remained of their family history. Apart from that, they had one set of clothes left on the clothesline.

I remembered the shock when my sister's house caught fire, sparked by faulty wiring in the loft. The only indication had been an early-morning power cut then a tendril of smoke drifting from the loft hatch. Flames took hold quickly and the fire brigade was on the scene within minutes, dousing the fire with powerful jets of water. Even so, the roof was gutted, the top floor a mess and the ground floor ruined through water damage.

There was no fire brigade for Rob and Lyn.

Lyn's voice took on an edge of fury as she described what happened the next day.

'This car turned up and this fella got out of the car with a case. Jehovah's Witnesses. He said, "Madam, we can see you're in a spot of bother. Would you like a prayer?" Well I went for him. I said, "Listen mate, if there is a f***ing God, he wasn't f***king looking after us yesterday so why don't you f*** off and take your f***ing paper with you."'

She repeated the blistering attack with such force more than ten years later that I almost felt sorry for the man.

Lyn laughed. 'After us he went to Kidston and he said to them, "That woman out there is bloody crazy." He was white as a ghost!'

A couple from the Uniting Frontier Church turned up three days later. 'How much do you need?' they asked, opening a cheque book. Lyn turned down the generous offer but she was grateful for their unconditional support. 'There was nothing about religion, they just wanted to see what they could do to help.'

The Jehovah's Witnesses stayed away. 'Next time they turned up was last year, first time since the fire. I've been wondering what I can do to offend them again.'

Lyn could laugh about the experience now but at the time it must have been horrific.

Insurance inspectors finally concluded rats had chewed through wiring in the ceiling above the bedroom, although the inevitable arson checks infuriated Lyn.

'I know he was only doing his job, but I lost it. I shouted at that inspector, "Why would I burn my own freakin' house down after we worked like dogs to get it? Get a brain you bastard!"'

Lyn's distress increased when ash samples collected for forensic testing went missing. 'I made the mistake of telling them we had a three-ounce nugget of gold in the ruins somewhere. I asked if I could please have the ash back.'

The ash was never seen again. 'We washed and sifted all the rubble and we accounted for every other scrap of gold. That three-ounce nugget would have been there, no question.'

A few weeks later Lyn read of a family who lost twins in a fire and she let go of her anger. 'We lost everything but no one was injured. We had each other, that was really all that mattered.'

They decided to build higher up the hill, well away from the danger of floods that had threatened the house twelve months earlier, and a transportable was ordered from a company in Townsville. That was the house we were now sitting in, with its open-plan layout, well-equipped kitchen and roomy bedrooms. It took twelve months to construct and install and while that was happening they lived in two dongas that survived the fire, turning what was left of the old house into a lick shed for cattle.

'You have to focus on the positive,' said Lyn. 'We didn't want anything fancy, just a house to live in.'

Rob's parents were connected to the grid and they lived just thirty kilometres away but the quote to install power to Lyn and Rob's new house was $280,000. They applied for a grant for solar energy instead, which gave them fifty per cent of the $82,000 they needed, and they scraped together the rest. 'We went without a lot to get it installed, but it was worth every cent.'

Thanks to the installation of solar panels they finally had a reliable source of power, backed up by a generator.

'That meant we could have ice-cream for the kids.'

Most of us take it for granted that we can grab an iceblock from the freezer. It's such a simple pleasure and I'd never thought about it before.

'Ice-cream defrosts too quickly and a generator's not reliable enough,' Lyn explained.

Knowing how traumatised the children were by the fire and its aftermath, Lyn promised them a new fridge-freezer, with an ice machine. 'We drove five hours to Charters Towers to buy new furniture they'd picked out for their rooms, plus a new fridge-freezer. Then we bought a tiny car freezer and we went to the supermarket and we drove home with ice-cream for them all.'

Lyn took a photo of her children sitting on the veranda, grinning. Kerri-Ann was sixteen, Ashley fifteen and Anna eleven when they had their first taste of ice-cream at home.

Next morning Lyn offered to take me on a tour of the property. Having confessed my ignorance of all things related to cattle, it felt easier asking the most basic questions, and I genuinely wanted to know more about the business side of things. Where did the cattle come from? What was a weaner? What did they eat? When did they sell them?

Lyn patiently talked me through the answers as we rattled along dusty tracks, slamming and shuddering over hard ground in a vehicle Lyn confessed she'd rolled and written off when she'd hit a kangaroo.

'We bought it back from the insurance company, cut it in half and turned it into a work ute.' Hopefully they kept the half with brakes.

Philip Hughes was right, the land at Gilberton was suitable for breeding cattle, not fattening, although close-up it didn't look that rough and it had more cover than I'd been expecting too. They had Brahman cross cattle (crossed with Romagnola, after trying European Charolais Braford, which were too soft; see I was listening) fed by grazing on grass that flourished after the rain each year, in shallow soil supporting small eucalypts, acacias, paperbark, spear grass, bloodwood and ironbark. Grass was supplemented by lick if the promised rain didn't arrive.

I could see what Lyn meant when she explained that the annual wet season was a vital part of the cycle at Gilberton, replenishing waterholes, dams and creeks. The landscape we drove through was a dust bowl, peppered with ubiquitous termite mounds, some of them several metres high.

'We love the wet season, you can see the country regenerate, even the trees look happy after the first downfall. The country breathes, it gives a big sigh and you can see the earth suck it up. Right now it's thirsty country. It's been eight years since the last really good drink.'

The big wet would normally start in February and run through until April, its arrival foreshadowed by flies that were sticky.

'Sticky?'

'More of 'em.'

Restless cattle that could smell the 'green pick' in the distance would often set off in search of it.

The wet season was when Lyn got all her chores done, sorting through photo albums, darning, mending, sewing and repairing machinery. Mail would be held at the local post office (which turned out to be 135 kilometres away) until there was an opportunity for it to be delivered. 'We've gone ten weeks without mail before.'

It wasn't unusual for the family to be trapped on the property for up to three months at a time, stuck on the other side of the raging Gilbert River. Three months trapped at home. I thought about the restless days I'd spent sitting in my studio in Sydney and calculated the most I'd ever managed without dashing to the local café for a cappuccino was a week. Who am I kidding? Three days.

In mid-December Lyn would stock up, ordering enough food and provisions to see them through until the following May. I had already marvelled that morning at the size of her pantry – it was larger than most people's lounge rooms – and now I knew why it was so big.

One year they tried a new supplier, phoning through the order weeks in advance. A huge pallet of goods was ready and waiting when they arrived to collect it from Townsville, everything neatly stacked and wrapped. They got home after the seven-hour drive across rough dirt roads to discover all the Sao biscuits – a vital standby – had been loaded onto the bottom of the pallet, underneath the tins.

'Jeez, there was nothing left of 'em but a load of dust!'

Lyn mentioned several times as we drove towards the ruined house that had since been rebuilt as a lick shed that Rob's parents were paid more for cattle fifty years ago than she and Rob were paid today, the reason being that most people don't buy from butchers anymore, they buy from supermarkets. Big chains can hold suppliers to ransom and control the price.

'There's only one meatworks in Townsville so there's no competition. We have to take whatever price we can get.'

If the promised rain failed to materialise, graziers would have to destock and hope for better conditions the following year. Budgeting for anticipated income from one year to the next depended entirely on the weather and on cattle prices.

Lyn laughed suddenly (possibly at the absurdity of some of my questions about profit). 'Prices can go through the roof or they can be down on the ground. We operate on minimum margins.'

Lyn and Rob could have put more cattle on their land. Others clearly had on some of the properties I'd driven through on my way to visit them and the result was obvious – a barren, blasted wilderness that looked more like the surface of the moon than Australian soil.

'We believe in stocking lightly. When your grass gets to a certain level you have to pull out the cattle, you don't just leave them there until the floorboards are showing through.'

They run 3000, and that includes the young ones. 'We could keep another thousand but we don't want to wreck the country,' said Lyn. 'You have to spell it. If you don't look after the land, it won't look after you.'

The paddocks I saw at Gilberton were well covered and for the most part thickly wooded.

'Every paddock gets a burn every three or four years, then it's locked up and spelled. It's worked for the past hundred years and we're not going to stop now.'

Something in the tone of her voice suggested not everyone would necessarily agree with their policy of regular burning, timed to coincide with the arrival of the wet season to avoid scorching the land. In Lyn's opinion burning kept ticks at bay, got rid of vermin, reduced woody weeds and tree thickening, and encouraged new grass. And who was I to argue with that?

'The kids would like to put a bit more Romagnola over the cattle, maybe put a bit more weight and muscle into them,' said Lyn, touching on how much trial and error went into the breeding program and the work that went into training the cattle to make them quiet and easy to handle. 'You get a name for yourself that way.'

Controlled mating, with all the bulls put in at a certain time, coupled with a ruthless approach to non-performers had gradually increased their calving rate from 550 to 800, but it had been a long, slow process. The prospect of drought could force them to destock and play havoc with those breeding plans.

Lyn came back to the stark fact that fifty years ago Rob's parents were getting more for their cattle than they did now.

'We've cut every corner we can, we don't employ people and still it can feel like we're going backwards.'

Feral pigs and dingoes were a constant threat to young livestock, although I was surprised to hear that chicken hawks and eagles – working in packs to attack young calves – did as much damage as dingoes. 'They're mongrels,' said Lyn, bluntly.

However much they planned, the unexpected was always just around the corner. Their mighty old workhorse of a truck, a 1993 Ford Louisville, broke down recently on the way to Charters Towers, just as it was getting dark. They were carting eighty head of cattle on a double-decker at the time, with another flat-top trailer behind, and the breakdown stranded them on a dirt road two hours from home.

'There was no point waiting for the RACQ to come out,' laughed Lyn (not that she'd have got through to them on a mobile anyway, there's no coverage out there).

The owners of nearby Welcome Downs Station came to the rescue and they helped unload the cattle, which were then mustered into their yards and given a drink while Lyn and Rob battled with the truck.

'We can normally fix most things but this was beyond us.'

In the end they had to remove a wheel, chain up the suspension, load up the cattle again and keep going. Four hours later they dropped the cattle in Charters Towers, drove another 140 kilometres to Townsville to drop the ailing truck, hired another, picked up gear in Charters Towers, drove home, unloaded the gear then drove another seven hours back to Townsville to return the hired truck, pick up their own repaired vehicle and drive home.

'Our eyes were hanging out by the end of it.'

I will never, ever complain again about sitting in traffic for ten minutes on my way to a pink slip inspection.

Dealing with a truck breakdown hundreds of kilometres away from a repair shop was just one of the challenges of living in such a remote spot. Unless good rains came, that truck would soon be pulling a semi-tanker, carting water from an open-cut pit a hundred foot deep to fill dams that were running dry.

Lyn had weather stats going back to the late 1800s and they supported her theory that they were in for a run of dry years.

'The big wet didn't come at all last year.'

If the drought looked like being a prolonged one, Rob and their son Ashley had plans to look for work elsewhere, leaving Lyn and Ashley's wife Camilla to look after the property . . . and plan for better times.

Lyn and her daughter-in-law were already thinking ahead, planning to build luxury cabins on the high ranges. The Gilberton

Outback Retreat would offer adventurous holidaymakers a taste of what it was like to live in cattle country: fly-fishing in the summer wet season and fossicking for gold in winter.

Until that ride around the property with Lyn I hadn't given any thought to the impact of environmental policy on people who lived and worked on the land. Lyn and Rob's concern for the environment and their determination not to overstock had a direct, negative effect on their income, forcing them to consider working elsewhere, even diversifying into tourism. How much money did my green credentials cost me? Very little. I'm a vegetarian but my partner's not so I shop for meat; I want choice and I want animal welfare; I want environmental protection and I want value for money. And I want people like Lyn and Rob to pay for it? It was a sobering thought.

Lyn pulled up outside the lick shed and jumped out of the car to greet Tokyo, a big white Brahman that was once a poddy calf needing a bit of extra care and attention.

'Andy? Andy!'

A bearded man in shorts, much like the old guy I'd spotted prospecting next to the caravan, appeared from behind a table strewn with bits of broken machinery. Andy had arrived as a prospector one day and Lyn had persuaded him to stay, helping on the property in return for food and a place to stay, as well as the elusive prospect of finding more gold.

The best time to prospect for gold is after a big wet. Prospectors still turn up at Gilberton hoping to strike it rich, some of them regulars who come back year after year. At Easter that year there were twenty-one camps dotted around the property, tucked in between the Gilbert River gums down on the flat.

It's not an impossible dream. When his city contemporaries were doing paper rounds to earn a bit of pocket money, Lyn's son Ashley was panning for gold. At sixteen he found enough to buy himself a motorbike. A year later, with the help of a metal detector and endless patient searching, he amassed a kilo. His uncle Morrie, a frequent visitor who normally lives in Brisbane, acted as bodyguard when Ashley walked into the Ainslie gold company on Creek Street in Brisbane, where his stash was checked and weighed. He'd found enough gold to buy a brand-new car.

We drove on and explored what little remained of the once-thriving Gilberton township, which in its heyday boasted seven pubs and a soft drinks factory. It had a courthouse, police barracks and a commissioner's residence, all of which were gradually left to decay after the short-lived gold rush collapsed in 1873 and white prospectors were chased out by Aborigines. The Martell family stayed.

Lyn's three children had six great-grandparents alive while they were growing up, right up until Anna was ten.

'They were pretty lucky, I reckon.'

Many of those great-grandparents were buried in one of the paddocks, in a cemetery that Lyn had started restoring eighteen months previously. There were no flash headstones, just mounds of earth with simple markers – rusting machinery and picket fences – standing under a clutch of trees. Some of the headstones had been washed away by flooding, and others had crumbled with the passage of time, but Lyn still knew where most of the relatives were buried.

'See that little mound? Grandma reckons that was Grandad's dog. They say that when Grandad died the dog sat at the front gate and cried. He didn't howl or bark, he cried.' Four days later the dog died and they buried the faithful companion next to his master.

She counted two steps, another two then four in the opposite direction.

'We think the Martell twins are buried here. They died back in the 1800s. There was a marker in the ground when I first came.'

The simple cemetery was above the flood line but water ran heavy in the wet and at some point it must have washed the Martell twins' marker away.

A plaque recorded all the people who had died at Gilberton and it was a sad reminder of their precarious, brutal and all too short lives: a rollcall of infants and newborns, as well as older children who died of influenza, scurvy, starvation and accidents, including Rosie, whose mother had been kicked by a horse and who died when she was a few hours old.

'That's just members of the family.'

Many more died unknown and largely forgotten, although Lyn has done what she can to honour their memories.

'That fella, he fought in Gallipoli,' she said, indicating the grave of a man who died in his early forties. He'd been a good friend of Rob's great-grandfather and he lived with the family for thirteen years. 'I told the government, the RSL, but they didn't want to know.' In the end, she tracked down a niece in Brisbane.

The biggest regret was that Rob's great-grandmother's body was missing from the cemetery. Flown out in the wet one year, she died in Cairns and her body couldn't be returned. 'Maybe one day, when we can afford it, we'll bring her back,' said Lyn, pulling weeds from the grave of another old prospector who had once lived with the family.

From the cemetery we went to look at a tiny stone fortress, barely six feet across, that had been built by the family in 1869. The great-grandmother who'd died in Cairns had been born in the fortress and Lyn had restored the building with help from an old prospector from Townsville.

'They holed up in here when the blackfellas attacked,' she said, knocking down a wasps' nest from under the eaves. 'Gilberton was the only town in Australia where the blacks were so hostile the Europeans fled. Only a few survivors stayed.'

I wasn't game to explore the tricky subject of Aboriginal history (about which I knew next to nothing anyway). That wasn't why I was there and, besides, my simplistic views had no place in a cemetery honouring the dead. I've always believed that there are many sides to any story and the truth probably lurked somewhere in the murky shadows between words like white and black, pioneer, settler, invasion, colonisation – words that held a wellspring of grief and rage. This was where it actually happened, where settlers came face to face with people who had lived on the land for generations.

I left the subject alone and we walked back to the car.

Sitting on the veranda later that day with morning smoko and homemade biscuits Lyn talked about how difficult it had been to try to teach her children when she herself couldn't read or write. It was a story that showed how fiercely she loved them and how determined she was that they should have access to the education she'd been denied.

'Oh, there were times I thought I had a screw loose getting married and having kids, I really did.'

With only a year between Kerri-Ann and Ashley, Lyn had decided to hold Kerri-Ann back so she and Ashley could start school together. In reality, she'd been putting off the moment when she would have to start teaching her children to read and write. It took less than three months for 28-year-old Lyn to realise it was never going to work.

'It was crazy to think I could teach my own kids, I could only cope with the most basic baby words – to, the, from, but – and I was doing my best but Kerri-Ann and Ashley hated toeing the line.'

Most days the lesson ended in tears, for everyone, and Lyn was forced to admit that she didn't have the first clue about teaching. Salvation came in the shape of Daphne Gear.

'She was an angel sent from heaven, the loveliest lady you could ever wish to meet.'

*

'Is she nice, Mumma?'

'No, she's horrid. She's got bad teeth and a big stick.'

'Why can't you teach us, Mumma?'

'It makes me cranky. And I'm too dumb.'

'She gonna live with us?'

'Yep.'

'What if we don't like her?'

'Tough.'

Lyn pulled up outside Cairns railway station and turned to face her children. It was May 1993, a hot steamy day, and she was at breaking point. In the back seat of the truck were six-year-old Kerri-Ann, five-year-old Ashley and five-month-old Anna.

'Now you remember what I said, you be polite to Mrs Gear,' Lyn said, addressing her two oldest children with a stern look. 'She's a mean old woman and if you don't do what she tells you, you'll be in big trouble.'

Kerri-Ann and Ashley pouted. 'We don't want a new teacher,' said Kerri-Ann, the oldest and by far the smartest.

'Well you're getting one. And if this doesn't work you're off to boarding school.'

'You wouldn't do that to us, Mumma!'

'Oh yes I would.'

Lyn got out and stood by the vehicle, waiting for the train to arrive from the Sunshine Coast. She loved her kids with a fierceness that sometimes took her breath away, and they were right, she would never dream of sending them away to school, but something had to change. The last week hadn't been so bad while the kids were on a School of the Air camp in Cairns, but when they got back to Gilberton she'd have to get out the school-books and start battling all over again.

Lyn heard the whistle of an approaching train and pulled herself together. There'd been plenty of cussing and swearing in the past eighteen months – cattle prices were shit and the cancer she thought she'd beaten had come back – but the only really dirty word in Lyn French's household was 'can't'. She always told her kids that if they wanted to do something badly enough they would find a way.

'You don't ever blame your background or the circumstances around you,' she would say. 'If you get a knockdown you stand up, shake yourself off and move on. Can't is a coward, can is a king.' She really hoped she could live up to her own advice.

Several minutes later an elegant woman with short grey hair and horn-rimmed glasses, who looked to be somewhere in her late sixties, stepped out of the station entrance carrying a suitcase.

'Hey kids, hop out, here she comes.'

Kerri-Ann and Ashley scrabbled out of the car and stood nervously beside their mum.

'You must be Lyn,' the woman said, extending her hand before looking down at the children. 'And this must be Kerri-Ann and Ashley.'

Five-year-old Ashley looked up at the smiling woman and he tugged on his mum's hand. 'Hey Mumma, she's not a mean old lady, she looks nice,' he said.

Lyn's heart sank. 'Oh jeez, out of the mouths of babes,' she muttered, and quickly added, 'I'm sorry, Mrs Gear, I can explain.'

The older woman laughed. 'Don't worry,' she said. 'I've heard worse, and please, call me Daphne.'

'Okay, Daphne. But listen up, kids, youse gotta call this lady Mrs Gear!'

*

Daphne Gear was a seventy-year-old volunteer teacher with Volunteers for Isolated Students' Education (VISE) whose Mary Poppins-like arrival rescued Lyn from the brink of despair. By the time they arrived home on that hot day in May 1993, Lyn was feeling more positive. Less than a week later, a huge weight had been lifted from her shoulders.

'Daphne Gear was a total stranger who came into our home and stayed for seven weeks. She introduced a set routine doing the abc every day and multiplication tables every morning. She was a godsend.'

Within a week the kids were in Daphne's room each morning, jumping into bed with her, although they stuck to the golden rule of never calling her Daphne. They settled on Mrs G.

An experienced teacher, she was a wise old soul who spotted Lyn's problem early on. Without making a fuss she introduced small tasks for Lyn to complete, little bits to read that would help her along.

'She was phenomenal, she wasn't down on me at all for not being able to read.' Two weeks later, Lyn opened up to the stranger in their midst. 'I just broke down one night, sobbing, and I told her my whole story.'

Daphne responded by telling Lyn she had lost a child.

'From that day we were really close. She was so good to me and the kids.'

Lyn absorbed everything she could about reading, writing and teaching. 'I spent ages with her, watching what she was doing and joining in with the lessons. I was like a funnel, sucking it all up. I knew what sort of mongrel kid I was and I wanted my kids to have an education.'

Their schoolroom was the kitchen table. 'I would put the books out in the morning and put them away again at night. It was pretty basic and feral but we were comfortable. Home is only what you make it.'

Rob wasn't too concerned about his wife's inability to read and write – ten years after their wedding he still hadn't finished reading the book he'd started when they got married – but Lyn was determined to improve. 'The more I learned, the more I wanted to learn. I loved reading.'

Daphne put systems and routines in place to help Lyn after she left, and Kerri-Ann was a bright child who often noticed when her mother was struggling. 'It's okay, Mum,' she would say. 'Grandad will be down soon, we can ask him.'

Rob's father would patiently work through whatever problem they were struggling to solve – often a math question – then he would demonstrate how he got the answer.

'Long division blew me away. Grandad Gus only got to grade three then he went droving but no calculator could beat him. He was unreal with figures.'

It wasn't easy sticking to a routine when the kids wanted to be out on the property with their father and their grandfather. 'It got so bad I even tied Ashley to the chair with a rope. Never mind about animal welfare, we were struggling with children welfare!'

Lyn laughed long and hard and I thanked my lucky stars that her kids were now all grown up and I wouldn't have to worry that the publication of this book might lead to calls for her prosecution.

'Ashley would hear Rob calling on the radio, saying a pump had broken and could we bring a spare part to this paddock or that paddock and – *ktchaw!* – he and Kerri-Ann would be out the door, over the shed and gone. I'd shout, "Get your reading books, take 'em with you!"'

The arrival of another VISE teacher each year helped focus their attention on the demands of schoolwork, but Lyn admitted that the competing demands of station life in a remote area often took precedence over rigid school timetables.

'Sometimes we just had to shut the schoolroom door and go mustering. They still did their studies but we couldn't afford to employ people so we had school holidays when it suited us.'

There was a defensive edge to her voice and I wondered if she was worried I might judge her? She needn't have been; I dropped out of teacher training just three weeks into the course, unable to cope with the rigidity of a centralised education system that forced children to fit into boxes and pass set exams.

But there were plenty of people who did take issue with Lyn's approach, including the then principal of the School of the Air.

'This male principal told me I couldn't teach my kids. He and I locked horns a few times. I wouldn't let the kids do calculator work. How were they meant to learn mental arithmetic if they had a calculator?'

Lyn would score a black line though the page on calculator work, get them to do 'a big mob of sums instead' then paste the sums in the book and send it back.

'It pissed them off and they'd make remarks that pissed me off.'

The growing antagonism came to a head when Kerri-Ann was in grade seven. Up until then she'd been a straight-A student, whose English compositions had all centred on horses and cattle. An exasperated teacher sent a note back with one of the essays about cattle, saying if Kerri-Ann wrote another story about a horse or a cow her work would be marked down.

'I lit up! I went to the principal and I said this is bullshit! Who cares if she always writes stories about horses and cattle, she's using her brain!'

From the age of three, Lyn's firstborn could identify a beast as a weaner and recognise it again as an adult. 'Kerri-Ann is phenomenal, she can read cattle.' Lyn believed her daughter's extraordinary connection with cattle was a gift she'd inherited from Rob's grandfather, and she saw no issue with her daughter writing stories about what she knew. 'We might have a thousand head of cattle in the yard and Kerri-Ann would know which calf belonged to which cow. She got it right every time.'

Even when they tried to trick her, Kerri-Ann saw through it. 'We would tell her we'd sold a particular cow to the meatworks and she would say, "No you haven't, she runs on that creek down there." And she was right!'

Lyn's defensive tone switched to downright contempt as she recalled the clashes with school policy.

Almost a year younger than his sister, and in the same grade for schoolwork, Ashley was struggling to keep up. 'I wanted to keep him back a year but the school said I couldn't do that.'

Lyn shook her head at the memory and narrowed her eyes. 'I said he is my child, I will do what I want!'

The arrival of a special-needs teacher from Amsterdam – a Dutch backpacker who wanted to spend time in the bush – averted a potential stand-off with the school. Diewerke Van Leeuwen, known to everyone as Dee, loved her stay at Gilberton so much that after six weeks of paid tuition she offered to stay on for another six weeks of unpaid work. Her arrival led to a marked improvement in Ashley's schoolwork.

When Ashley discovered Dee could ride a motorbike, he gave his Dutch tutor his full attention and knuckled down to study. Outside school hours he went all-out to impress her with his undoubted skills. According to Lyn, nobody could ride a motor-bike like Ashley.

'He could be chasing a beast at full speed then come across a barbed wire fence. If the cow jumped the fence, Ashley would throw the bike on its side, slide it under the fence, leap over the fence then grab the motorbike and be back chasing the cow before you could blink.'

It sounded terrifying and I was quietly relieved when Lyn admitted that her heart still missed a beat each time Ashley flew past the homestead on his back wheels, grinning at her.

The arrival of computers didn't make life any easier in the schoolroom, if anything it only added to Lyn's troubles. With a connection powered by a generator, downloads were slow and intermittent and the children's work took far too long to send. Lyn's answer to the noise and fury of everyone's growing frustration was simple and effective.

'I introduced a money bottle and everyone got two dollars at the start of each week. Anyone who lost it and started screaming "I can't do this!" – including me – had to pay the others twenty cents. If nobody yelled that week they all got an extra two dollars.'

It would be a mistake to think the lessons were all a battle. Lyn was a talented pianist, who, until she ran away from home, had dreamt of one day becoming a music teacher, so the one lesson they all loved was music. She was confident of her abilities to help Kerri-Ann on tenor sax, Ashley on drums and Anna on flute, supplemented by weekly lessons on School of the Air. When Her Majesty Queen Elizabeth asked for the School of the Air band to play during her tour of Australia in 2000, all three children were invited to join the band.

'We drove to Cairns a couple of times for music camps but most of the practice was over the air.'

After the thrill of performing live in front of the queen the band went on tour, playing along the east coast and in outback towns such as Emerald, Longreach and Hughenden. Amid all the excitement, one of the most memorable performances was in Eventide, where the children's great-grandparents were living in a retirement home.

'They got to hear the children play for them six months before they died.'

Lyn went to extraordinary lengths to pursue her dream of educating her children. When she discovered that a school-based traineeship in cattle production through Rural Industry Training and Extension (RITE) wasn't available to distance education students, she fought to get Ashley and Kerri-Ann into the course.

'They were the first distance education kids in Queensland to do it and they completed a two-year course in eight months. Ashley won trainee of the year,' she said proudly.

Unknown to anyone, Anna decided to do the course as well. She was only eleven at the time, but she'd been out in the bush with her father and grandfather practically since the day she was born. 'Rob would take a nappy bag out with them and away she'd go.' Without telling her parents, Anna wrote to the CEO of the program, asking if she could do the course:

I am eleven years old. I am wondering how old I have to be before I can join RITE . . . I do school at home on our station. I can do any of the station jobs such as mustering, fencing, doing lick runs . . . I am at present breaking in my first horse with my grandad. I want to stay on the land and I think RITE would help me. I hope to hear from you soon.
Yours faithfully, Anna French.

A special board meeting was held to discuss the request.

'The first we knew of it was when they sent us a copy of her letter.'

The board decided that Anna should be allowed to take the course, a decision that was overturned when the education department refused to fund such a young participant, arguing she wasn't old enough to tackle branding and earmarking.

'She'd already been doing it for years!'

Lyn took up the fight. She appealed, the matter went to court and Anna was asked to give video evidence. 'I was so proud of that girl,' said Lyn, her voice tight with emotion.

Anna faced the camera to deliver her argument.

'It's all right for youse people with shiny suits and slippery biros, youse don't have any idea what it's like,' she said. 'Youse come out here and work with me then you'll see.'

The challenge was turned down and the appeal refused but Anna wouldn't be beaten. 'We'll find another way, Mum,' she said.

There were some in the education department who questioned Lyn's judgment in pursuing the matter, which only strengthened her resolve. 'I said, if you've got a kid who loves football, you take them to practice every week and you help them pursue their dream. Well this is my daughter's dream and it will happen, whether you agree or not.'

And happen it did (which didn't come as any surprise to me by that stage). Anna wrote to Connellan Airways Trust, a fund established to promote the education of young people in remote areas of outback Australia, and they offered $3000 towards the cost of the course. Lyn and Rob paid the remainder and in 2003 Anna French became the youngest participant ever to be awarded Certificate II in beef cattle production. At fourteen she went on to complete an online diploma in beef cattle production through Dalby Agricultural College, again with help from Connellan. By fighting for her children to undertake a school-based traineeship through distance education, Lyn paved the way for other rural students to follow a similar path.

'Like I always told my kids, can't is a dirty word.'

Lyn was pragmatic about any suggestion that her children might have missed out on opportunities. 'Their schoolroom was the size of Belgium,' she said, referring to the vast Australian Outback.

'You get a lot of criticism, people say you're depriving your kids of this and that and they need an education, and I agree, but I see a lot of people who have pushed their kids away and they've lost the connection with them. All our kids are on the land now and people are saying you are so lucky.'

Her face betrayed what she thought of the suggestion that luck had anything to do with the way she raised her children. 'They were taught to believe in themselves and to respect others. They learnt that hard work brings rewards and they learnt to strive to achieve their dreams.'

None of the children wanted to stay at school beyond grade ten, although all were given the opportunity. Anna briefly attended boarding school in Charters Towers – something she'd wanted to do because all of her friends were doing it – but she didn't enjoy it. After four weeks of nightly sobbing phone calls, Ashley stepped in.

'Right, Mum and Dad, this is the deal, if you don't go and get Anna and bring her home, I will!'

Lyn has shown her children that education doesn't have to stop when they leave school. Learning to read and write in her thirties gave her the confidence to pursue opportunities she would never have thought possible. Lyn spent eight years as a volunteer co-ordinator and eleven (and still counting) as a blue card administrative officer for VISE; she has been a long-standing director on the board of RITE; a volunteer firefighter; a co-ordinator for Drought Aid in north Queensland; and she spent six years as a member of the federal council for the Isolated Children's and Parents' Association – an assignment which regularly took her from Queensland to Canberra.

'The first time I went to Canberra it was freezing. I wore so many clothes to keep out the cold I must have looked like a wobbling penguin.'

I was filled with admiration for Lyn. Talking to federal politicians (not many of whom strike me as approachable) must have been a daunting prospect for someone who only a few short years before had been unable to read or write.

'I had a job to do and that was to educate politicians about the challenges facing rural and remote kids and parents. The biggest challenge was having to wear shoes!'

Lyn lobbied politicians on behalf of causes dear to her heart, including home-tutor payments for mothers who were forced to educate their children at home. Members of the federal council are still lobbying for that allowance, although it's hard to imagine anyone doing it with more passion than Lyn.

Lyn has since studied for a postgraduate diploma in land management through Rangelands Australia and she was recently awarded a scholarship from the foundation running the Australian Rural Leadership Program. At the time of our interview she was part-way through a two-year ARLP course.

Lyn never forgot the advice of an old Aboriginal fellow she met when she ran away from home all those years ago. He told her, 'Take whatever opportunity comes your way. Grab it and run with it because you never know when in life it might help you. And never be bitter. Place the mistakes under your feet and use them as stepping stones to rise above them.' He could have been describing the life that lay ahead for her.

Lyn and Rob's children are adults now and they all work on the land. At the time of writing, their youngest, Anna, was contract mustering up in the Northern Territory, happily settled with her partner, Joe, who survived the inevitable phone call from Lyn. 'She's my baby and if you don't treat her right you won't just have me to deal with, her dad's a big fella and her brother won't be far behind.'

Lyn's protective attitude was understandable when I learnt of the accidents Anna had survived. At seven, she was thrown from a horse while mustering in a high paddock fifteen kilometres away from the house (anyone who questions why a seven-year-old would be mustering should remember this girl could ride a horse when she was three). Anna suffered a broken pelvis and a compound fracture of her leg. They strapped the leg with a makeshift splint of reins and saddle blankets and, by the time they got back to the house, Anna had gone into shock. The Flying Doctor airlifted Anna to Townsville, where she spent three months in traction followed by another seven weeks in a full body cast. When she was eventually released from the cast she had to learn to walk again, and the first thing she wanted to do was get back on a horse. Nine years later Anna fell off a motorbike and almost lost an eye. They've all suffered broken bones at one time or another.

Kerri-Ann's abiding love of cattle grew stronger as she grew older. At eighteen she started her own stud – Spinnaker Brahman – and went on to manage a successful business with 8000 cattle on Glenmore Station, near Einasleigh, for a family company.

'She won't allow anyone to mistreat or abuse her cattle. She's a chip off the old block and a bloody good cattlewoman. She can read a beast and not many people can do that anymore.'

Ashley and his wife Camilla settled into a house about a kilo-metre away from Rob and Lyn, opposite the site of an original homestead that was washed away in the floods of 1918. They live far enough away to satisfy Lyn's golden rule – never live so close to your in-laws that you can see their washing hanging on the line – but close enough that Lyn can see her grandson on a regular basis.

Three-year-old Robert French represented the seventh gener-ation of the French family to live at Gilberton and it was clear from everything Lyn had told me that it was getting harder and harder to make a living from cattle. Would he be the last, I wondered?

'Hopefully it will continue,' she said. The prospect of a luxury holiday retreat being built above the Gilbert River would certainly help.

The isolation of living in such a remote spot didn't seem to bother Lyn. Even seven hours inland, surrounded by pristine bushland in such a remote spot, she still sometimes craved solitude.

'Your husband is your workmate and your best friend, and you rely completely on each other. After twenty-eight years I some-times want to say back off a bit and give me some space.' If that ever happens she gets on her motorbike and drives off.

'But no matter how hard it gets out here, I'd hate to live in town. It would drive me nuts.'

Lyn slept on a lounge chair recently when Rob's snoring kept her awake and she tackled him in the morning, grumpy through lack of sleep. 'I'm going to divorce you, you bastard, you won't stop snoring!'

'Where are you going to go?' the shy giant asked.

'I don't know.'

'Well, when you figure it out, can I come too?'

Lyn laughed at the memory. 'This is what I know and this is what I love, including him! What was I thinking of all those years ago saying I never wanted to get married or have children?'

Whatever love Lyn may have lacked in childhood, she has more than made up for it in adulthood.

'We're a close family and the kids are all good mates, but it's not something that just happens. You have to teach them to be open and honest and true to themselves. They've had their moments, like any siblings, but if anything happened they would be there for each other at the drop of a hat. And how many children get to work with their father and their grandfather?'

None that I knew of.

'We all really appreciate having the grandparents around, and they cherished having the kids here when they were growing up. We're all so fortunate. My family is my world. Without them and Rob, life wouldn't be worth living.'

Jo and Dave Fulwood

'Yarrandale'
Cunderdin, 160 kilometres
east of Perth, Western Australia

Jo and Dave Fulwood live two hours inland from Perth, slap-bang in the middle of the central wheatbelt area that stretches across a vast swathe of Western Australia. I visited them when paddocks of flowering canola, wheat and barley were shouting bursts of colour under a cloudless sky.

I'd heard that Jo and Dave were both from fourth-generation farming families and I knew they had three children. I also knew they'd been forced to call on the Flying Doctor for help when their daughter suffered a serious bout of pneumonia, but that was all I knew.

Jo was warm, chatty and engaging on the phone. She surprised me by admitting that neither she nor Dave had shown any desire to follow in the family tradition of farming when they left college. Jo had studied media and communications and they'd met in Melbourne, where their lives revolved around cafes, restaurants and travel.

A family crisis had prompted the move back to Perth and less than three years later they moved onto the farm at Cunderdin, where Dave was in his element from day one. He loved living and working on the farm and Jo hated it. She spent all her spare time plotting how to escape. Ten years on and three children later, Jo wouldn't dream of living anywhere else.

There had to be a story in there somewhere.

It turned out that Jo's experience of life on the land was different from every other woman I spoke to; she had no desire to be a hands-on farmer and she played no part in the running of the business. Jo's life centred around community, or in her case, the lack of it – Cunderdin had no childcare facilities, effectively trapping her at home with three children under the age of two. Serious healthcare issues with one of the twins exacerbated Jo's sense of isolation.

Not one to walk away from a challenge, Jo tried to galvanise support in the local community for a childcare centre, and in an epic David and Goliath battle the publicity-shy Jo appeared on radio, television and the front page of local newspapers. She fronted a campaign more challenging than any she'd run behind the scenes.

Thanks to that campaign – and to the grit and determination that characterise so many women in rural and remote areas – Jo found her place in the community.

Green and gold

I didn't want to interview Jo Fulwood any more than she wanted to be interviewed. The problem from Jo's perspective was that she didn't think she had a story to tell but I knew she did. My problem was that Jo was a staff writer for the prestigious *Countryman* newspaper in Western Australia, and a farmer's wife to boot, so not only would she be extremely knowledgeable about all aspects of farming, she would also know everything there was to know about interview techniques. At the age of fifty-four, with only intermittent stints as a journalist, my interview skills were as rusty as an old plough that had been left in a paddock for several long hot summers. I resolved to let the tape recorder run and pick up what I could (and keep the farming similes to a minimum).

There was only one road from Perth to Cunderdin and that was the highway that eventually led to Kalgoorlie. The monster trucks I passed would have made a child's eyes widen with disbelief. The bloated giants were visible from kilometres away, carting heavy slabs of equipment that hung over the sides – an improbable march of machinery like something out of *Lord of the Rings*.

The landscape though was surprisingly beautiful. I'm not sure what I was expecting on the drive out of Perth, certainly not the wilder version of Tasmania that greeted me, with dense forests and mobs of sheep grazing in green fields (oops, paddocks).

The further inland I drove the more crops appeared. There was no sign of the pivot irrigation systems I now had some knowledge of (so much for my opening topic of conversation) although I did spot what I took to be a large water pipeline following the road.

Approaching Cunderdin two hours later was like driving through a film set. Paddocks full of wheat, barley and yellow canola stretched into the distance, shifting in the breeze and saturating the landscape in a technicolour glow of green and gold that shimmered under a cobalt blue sky. I felt a sudden impulse to wind down the window and shout 'Aussie, Aussie, Aussie, Oi, Oi, Oi!'. Glimpses of red earth under the flourishing crops reinforced the magic. Was Dorothy in there somewhere, skipping along the beaten earth on her way to visit the wizard?

I pulled myself together (what would she *think*) and turned off the highway, following Jo's directions until I found the commemorative rock that marked the entrance to Yarrandale, the property that had been in Dave Fulwood's family since 1910. I passed a rundown transportable house on my way up the hill and I parked in front of a gleaming modern house, shaded by trees and surrounded by flowering canola.

'Come back in February when it's forty-five degrees and it hasn't rained for six months and the landscape is a barren desert. It's not so pretty then,' said Jo bluntly when I enthused about the view. (Thank God I didn't mention Dorothy.)

We were standing in an open-plan area at the centre of her house – kitchen to one side, dining area in front of us, big open fireplace behind – looking across a well-tended garden towards paddocks planted with waist-high crops of flowering canola. It was hard to imagine the area as anything other than productive.

'Dave is chair of Nuffield Western Australia and as part of his work he met some farmers in England. One of them told him that by world standards anywhere with less than 300 millimetres of rainfall a year is considered desert. Our long-term average is just under 300 so we are officially by world standards living in the desert.'

This wasn't like any desert I'd ever seen.

'You've arrived at the perfect time. If you were to interview me in February, after a run of 45-degree days when we won't have seen a blade of green from October until May, it would be a different story.'

Having lived in Broken Hill for several years, I knew what Jo was talking about. I also knew I would always think of the central wheatbelt area of Western Australia as a gloriously productive, breathtakingly beautiful place.

'I love it now though, this is my favourite time of the year,' Jo conceded.

So why had she 'hated it' when she first moved to Cunderdin, I wondered. Hate was a strong word and it was far too soon in the interview to start asking such direct questions.

We talked more about farming, family and work (or rather Jo talked and I listened, with growing respect for her knowledge). She was entertaining company.

Jo wouldn't fit any stereotypical idea of a farmer's wife, despite the fact that she grew up on a sheep farm three hours south of Perth. A slight woman in a pink jumper, tracksuit pants and ugg boots, with long blonde hair drawn back into a ponytail, the attractive mother of three looked far younger than thirty-nine. I was surprised when she laughingly admitted that she had no idea how many paddocks they had. That didn't mean she was ignorant about agriculture though – far from it. Jo was extremely well informed and her position as a journalist with the *Countryman* newspaper gave her a unique insight into the lives of farmers and graziers right across Western Australia.

Their 4000 hectares at Yarrandale were devoted entirely to cropping, with just a few 'lawnmower' sheep. Behind the house, far enough away so her three children wouldn't be tempted to use it as a playground, the farm bristled with heavy machinery run by state-of-the-art GPS to regulate the amount of fertiliser, seed and herbicide each parcel of land required.

'I'm not machinery minded,' said Jo, cheerfully. 'And I'll probably get shot down in flames for saying this but I think it's more of a man's world.'

Such candour was refreshing and I sensed us both beginning to relax. I relaxed even more when Jo admitted she hadn't been working as a journalist for that long.

'I'm lucky and grateful that the editor took a chance on me.'

Jo's working life revolved around story ideas and newspaper deadlines and she still pinched herself that she'd been offered a job on a newspaper which allowed her to work from home, although the more I heard the more I realised there can have been few people better qualified.

Jo was from a fourth-generation sheep farming family in southern Western Australia. Her father, Des O'Connell, had been forced to sacrifice his plans to go to university when his own father, Jo's grandfather, fell seriously ill.

'Dad was in year twelve at the time and there wasn't enough money to send him to university. He had to go home and take over the family farm. It was either that or it would have been sold.'

Des helped his middle brother through university and had the satisfaction of seeing his younger brother go as well, but his own education was sacrificed in the process. Both of Jo's uncles were now highly regarded specialists in the fields of biochemistry and microbiology.

'Dad would have thrived at university if he'd had the chance.'

With a mother who was a teacher and a father who wanted his children to benefit from the opportunities he had missed, Jo and her siblings were destined for further education. She had no desire to be a farmer so Jo studied media and public relations at Edith Cowan University.

After a formative year in an ad agency followed by a four-year stint as communications director for the WA Farmers Federation, Jo took a job with the Australian Wheat Board in Melbourne, a city she fell in love with from the moment she arrived.

'I was a farm girl from Western Australia and I thought Melbourne was *so* exciting.'

Jo was the kind of person who made friends slowly and cautiously. Her natural inclination was to stay in the background but the energy and pace of her new urban environment swept that caution aside. With new people arriving in the city all the time she found it easier than she'd expected to make friends. She had a high disposable income and she lived in a rented house in Port Melbourne – right behind the main Bay Street café and restaurant strip. Jo spread her wings and flourished, picturing a cosmopolitan future stretching ahead of her.

Dave's background and upbringing were remarkably similar to Jo's. Like Jo, he was from a fourth-generation Western Australian farming family, and, like Jo, he had two siblings. Dave was a sociable, outgoing man, based in Western Australia but working and travelling across Australia, and not long after they met he took a job with Goodman Fielder in Melbourne.

'We loved the Melbourne lifestyle and the café culture. I assumed that somewhere along the line we would probably end up living and working in London.'

It was a long way from cosmopolitan London to rural Cunderdin but before I could probe any further our conversation was interrupted by her two sons, nine-year-old Hamish and seven-year-old Seb, who exploded into the kitchen in search of toast.

I took the opportunity to glance around. Jo's kitchen wouldn't have looked out of place on the pages of a glossy magazine. Its clutch of modern appliances, including two coffee machines, a waffle maker, blender, juicer, mixer and all-singing, all-dancing Thermomix, suggested someone time-poor who loved to cook. (Never judge a book by its cover – they were the product of a husband who loved to shop for gadgets.)

When the conversation resumed we were back on farming and Jo had a wealth of knowledge about the agronomic issues they faced, from non-wetting soils to resistant weeds.

'But frost is the thing that keeps us awake at night.'

More than drought, which to some extent could be foreseen and action taken to mitigate its effects, frost was their biggest worry. Frost could wipe out an otherwise profitable harvest in a matter of hours.

Jo explained that if an overnight frost happened when wheat was in flower it would sterilise the grain and, come harvest time, the fully grown plant would have an empty head. A stem frost could happen later, once the wheat had finished flowering and the grain was sitting in the head waiting to ripen.

'Frost at that stage can freeze the stem of the wheat plant, which cuts off the nutrients. If that happens the wheat will shrivel up and become a hollow shell.'

So for all their farm's high-tech equipment and modern farming techniques – yield mapping and precise tramline farming made Western Australian farmers some of the most efficient in the world – they were still at the mercy of the weather.

'A lot of farmers are looking at mould board ploughing. It seems to help with frost *and* weed control.'

(Stick with me on this one, it gets seriously interesting.) In dryland agricultural areas like Cunderdin, where they relied solely on winter rainfall to produce crops, any summer rain caused weeds to germinate. High levels of chemical control had traditionally been the answer, which had led to serious resistant weed issues.

As a mother of young children Jo had always felt uneasy about their farm's reliance on chemical weed control.

'Many is the time I've been pegging out washing when a crop sprayer has flown overhead, forcing me indoors.'

Jo had often wondered if the use of pesticides might explain the high incidence of cancer locally, citing several neighbours who had succumbed to the disease, including Dave's mum. Her good friend Nat was about to undergo a double mastectomy and her own mother had just finished her third bout of chemo.

'My parents are clean-living country folk. Mum never smoked, she doesn't drink and they stopped using plastics in the microwave and shampoos with sulphates long before it was fashionable. There was no incidence of cancer in the family so you have to wonder about environmental factors.'

The 'mould board ploughing' Jo mentioned was a centuries-old farming method that originated in England. It was a non-chemical way of reducing weed and seed populations by inverting the top thirty to forty centimetres of non-wettable soil, burying weeds so deep they couldn't germinate. At the same time, the process brought more wettable soil to the surface.

Dave had been at university with someone who now imported the ploughs from England, and they were trialling it on Yarrandale with positive early results. The technique was introduced more for weed control than frost protection, but it seemed to work for both.

Anything that cut down the use of chemicals got the thumbs-up from me. A diagnosis of multiple chemical sensitivity had forced me to rid my own house of most chemical-based products.

It was time to bundle the boys into the car to go and collect Seb's twin sister, Annie, from dance class.

It was dark by the time we got back. The concrete slab floor had been warmed by the sun earlier in the day and had slowly released its heat. It was cool to the touch now, although the well-insulated house didn't feel cold. Jo struck a match and lit a pile of kindling under the logs in the central fireplace. The fire spluttered and took hold, dispelling the cold winter night that pressed against the uncurtained windows.

It was an extremely spacious and comfortable house – brick-built, single storey with an angled pagoda-style roof that took advantage of the winter sun and kept out the blistering summer sun. The fire was their only source of heat, and the house didn't have air conditioning either. Given the type of summer heat Jo said they experienced in the wheatbelt area that struck me as a glaring omission. I wish I'd kept my opinions to myself. A shocking drought the year before they built the house in 2009 had devastated their expected yield and left them with no money for non-essentials, like heating or air conditioning.

The house was designed around an outdoor room, which had been closed off that night to keep out the cold. The room relied on a large fan and misters to combat the heat in summer. I felt for Jo, remembering the exhaustion of 45-degree days in Broken Hill, but she was philosophical about it. 'We couldn't afford double glazing either, but you know I've never been as cold here as I was in Melbourne.'

There was that mention of Melbourne again. It kept cropping up. Suddenly seven-year-old Sebastian burst into the open-plan living area, naked apart from a towel, giggling as he raced around the room.

'Seb, go and put some clothes on and calm down,' Jo urged.

She had little hope of being heard. Hamish was in hot pursuit and he raced Seb around the dining table as Annie sat patiently at the breakfast bar waiting for her mum to listen to her read, fingers pressed under a word she was struggling to pronounce. Jo pulled up a stool.

'That's right, Annie, se . . . gu . . . sea gull. Well done, sweetheart.'

Seb tore along the corridor and there was the sound of a door being slammed shut then locked.

'Have you had your shower?' Jo called.

There was no answer apart from a muffled giggle and Jo turned her attention to dinner while I sat and helped Annie with her reading.

Dave was away, the fourth time in as many weeks that he'd had to visit Sydney, Melbourne, Adelaide or Perth for yet another meeting of the ministerial task force he sat on. The *Countryman* was on deadline. Jo had finished her latest interview, written the copy and filed the story but the photo she needed hadn't come

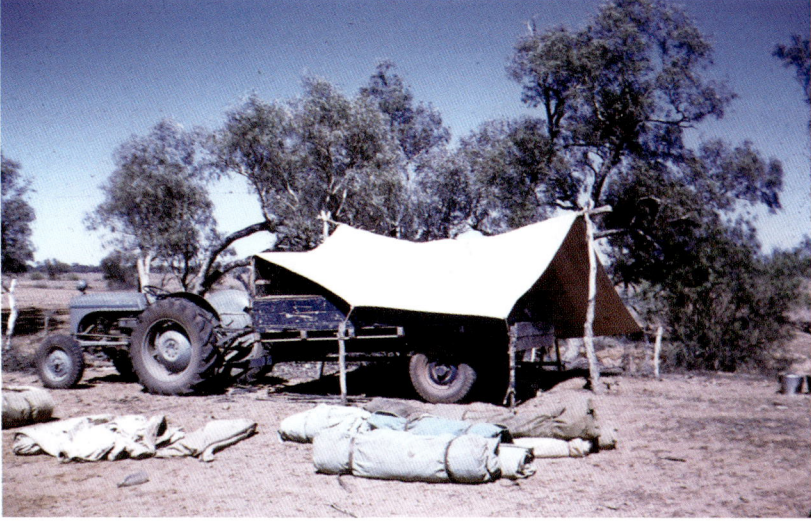

The kitchens Adele Hughes encountered after she got married were nothing like the Melbourne kitchens she grew up with. Adele's husband, Philip, worked as head stockman for two years at Koolatah, 400 kilometres west of Cairns, where the stock camp kitchen (pictured here in the 1970s) was equipped with swags for overnight accommodation.

Adele Hughes (centre) once considered opening a gourmet cheese shop in Melbourne. Instead she married a stockman and raised her children on cattle stations. Her father, Murray Crow (left), visited Heatherlea in Queensland in 2012 to see his newborn great-grandson, Willliam, cradled in the arms of proud father, Lachlan (right).

Three generations of Hughes cattlemen gathered at Dulacca Downs in 2011. Philip Hughes (second from left) and his father, Bill (second from right), followed a long line of Welsh ancestors who migrated to Australia in the 1940s and settled on the land. Five generations later, Philip and his sons, Lachlan (left) and Alister (right), work in the family business together.

Fifteen-month-old Hamish is dwarfed by the machinery on Dave and Jo Fulwood's farm in the central wheat belt of Western Australia. He's pictured tottering through the stubble in early 2006.

Dani McCreery

By contrast with her husband, Dave, journalist Jo Fulwood hated her new life in Cunderdin, Western Australia, and plotted ways to escape. Now she wouldn't dream of raising their children anywhere else. Hamish (eight), Jo, twins Annie and Seb (six) and Dave love living on their farm in Cunderdin.

Kate Raston

Kate Raston

Jo is a passionate advocate for greater recognition of the importance of agriculture and the role of women in rural areas.

Jo knows she would constantly be apologising for the noise if her kids lived in a city. Hamish, Annie and Seb have all the space they need to run wild.

Rick Prevost (right) admits he wasn't a natural farmer. His daughter Virginia (centre) inherited the farming gene from her grandmother and she took over her share of the Epping Forest farm, Tasmania, at twenty-three. She now commutes from husband Steve Chilcott's dairy farm in the Meander Valley, where they live with their two young children, Georgie and Henry.

Virginia credits a year in America with making her the successful farmer she is today. She's pictured here in 2001, age nineteen, at the American National Sheep Show.

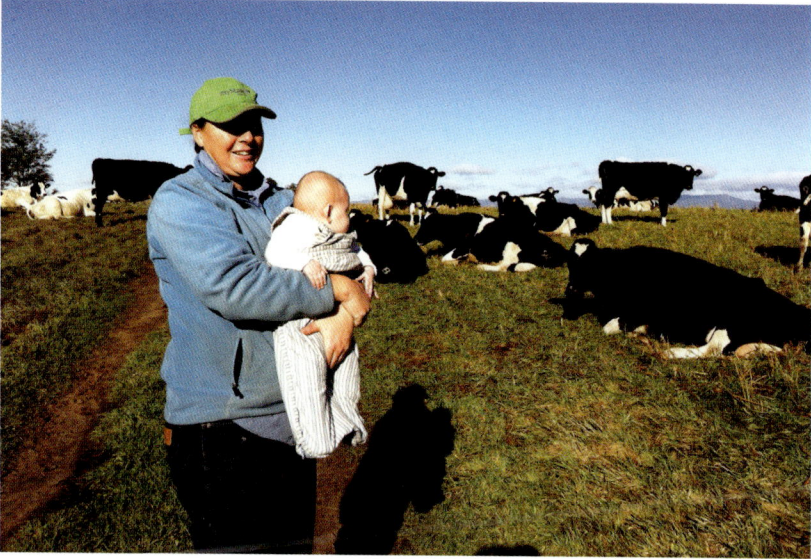

Georgie has probably spent as much time in a paddock as she has in a pram. At sixteen months old, she and mum Virginia Chilcott survey contented cows on an agistment in Epping Forest.

Lyn French ran away from home at fourteen, unable to read or write, vowing never to get married or have children. She's pictured second from left, sitting on the rails of the cattle yards at Gilberton Station, Queensland, with husband Rob and children Anna, Kerri-Ann and Ashley, who were all home schooled.

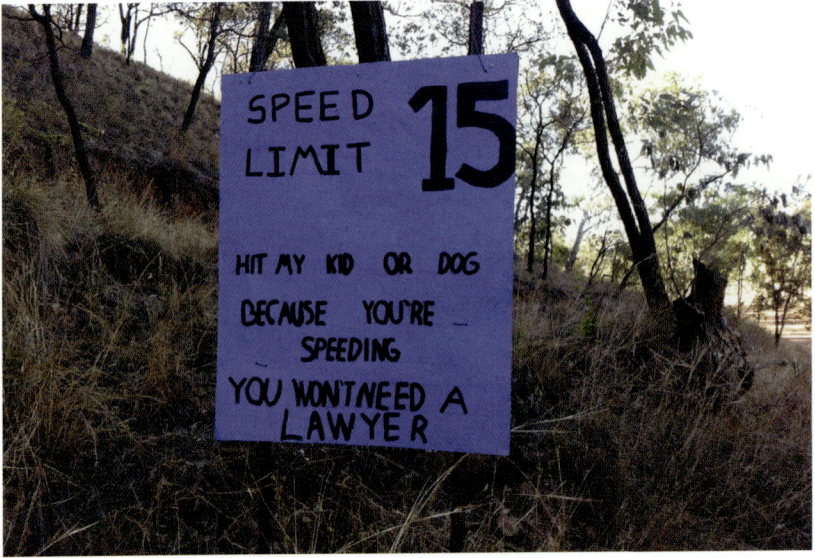

This handmade sign greets visitors to Gilberton Station. After a seven-hour drive inland from Townsville, on largely dirt roads, I wasn't taking any chances and took my foot off the accelerator.

Gilberton Station has been in the French/Martell family since Rob French's ancestors arrived in 1869, drawn by the short-lived gold rush that saw them establish a butcher's shop and take up 88,000 acres to run cattle.

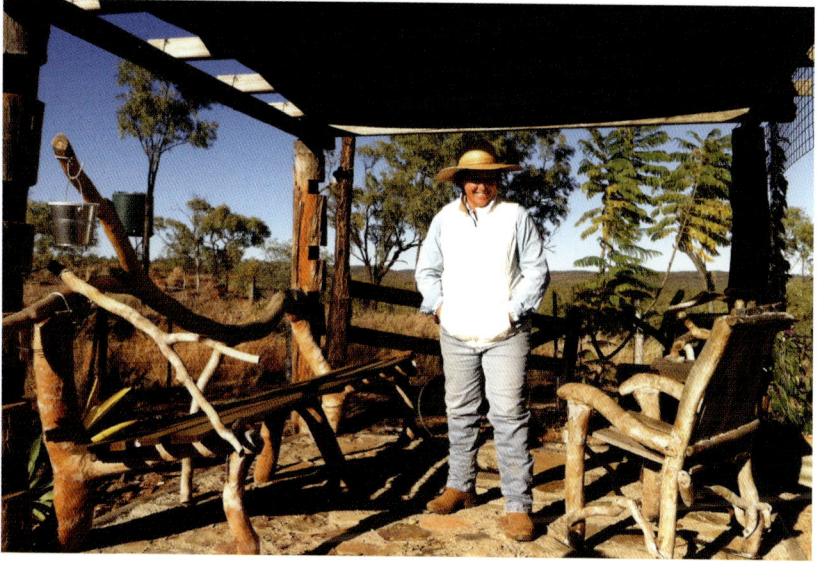

Gilberton is in the rocky high country of northern Queensland and this rustic lookout behind the station is a favourite spot for a beer at the end of a long hot day.

Little Robert French is the seventh generation of the French/Martell family to live at Gilberton Station. Here he helps his grandfather, Rob, make a holding paddock on the property in 2013.

The remote Durham Downs cattle station in south west Queensland covers an area of 8910 square kilometres and is sometimes referred to as the 'jewel in the crown' of the Kidman cattle empire.

There's no shortage of space for Jon and Michelle's boys to play outside on Durham Downs yet, like most kids, Joel, George, Keegan and Will often have to be prised away from their computers.

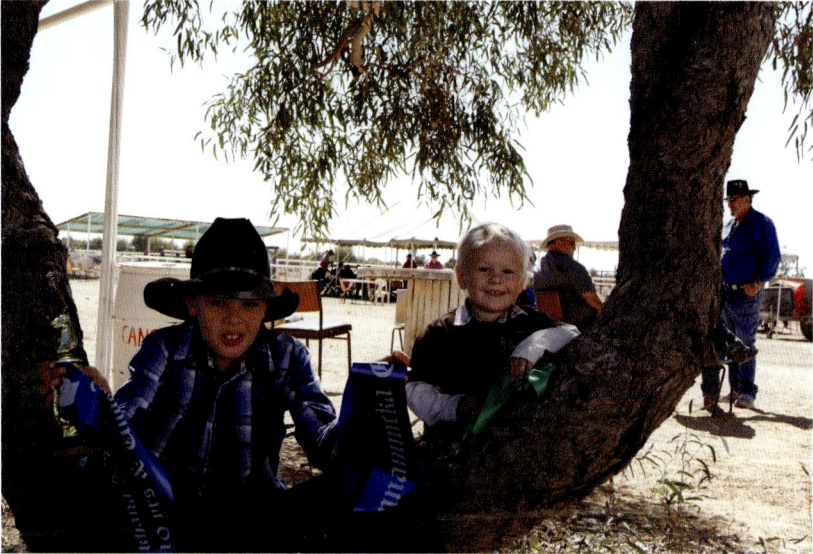

Will and his older brother, Keegan, took out several prizes at the Innamincka Gymkhana in 2011, an annual event that gives the boys a chance to socialise with other families from neighbouring stations.

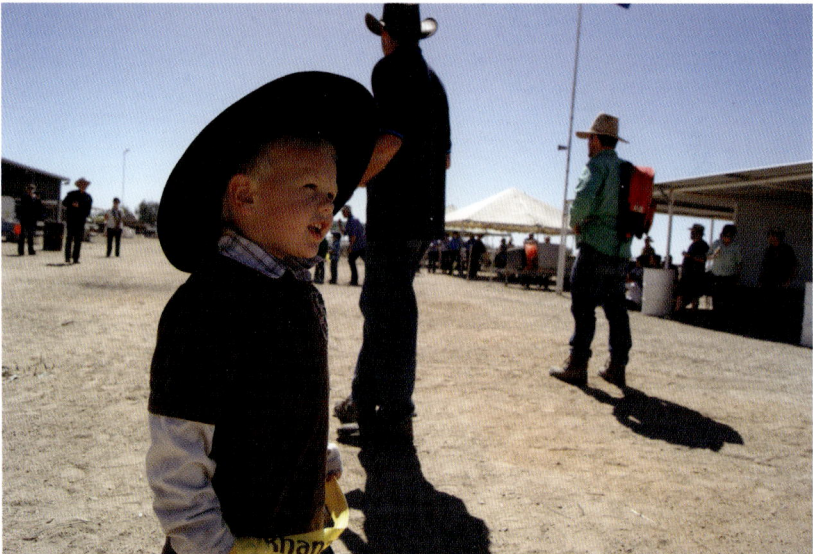

The Royal Flying Doctor Service airlifted the youngest of Michelle's four boys from Durham Downs to Adelaide after he suffered a serious accident in the meat house. There was little evidence of Will's mishap at the Innamincka Gymkhana in 2011.

Dogs are a big part of life for Michelle and Jon's family, from working cattle dogs to treasured family pets. Will cradles young pup Jess, hoping to keep her safe from the multiple hazards of dingo baiting, snake bites and rat poison.

Lorraine Kath

Michelle's love of horses is part of the reason why she enjoys living and working on the remote cattle station of Durham Downs. She demonstrated her horse riding skills at a camp draft competition in Birdsville in 2013.

The Britnells built their Woolsthorpe dairy themselves, saving thousands of dollars in the process. Roma and Glenn are pictured with their daughter, Tessa, and two of their three sons, Vincent and Austin.

Young calves learn how to drink from a bottle by first sucking on someone's fingers. At Roma and Glenn Britnell's dairy in Victoria, all four children were expected to help with the feeding routine, including their youngest and only daughter, Tessa. She looks happy enough in this photo, taken in 2010 when she was seven, but Tessa took a while to accept her fingers would have to be covered in cow slobber.

Roma Britnell's family surprised her with a professional photo shoot for Mother's Day in 2011, taken on the farm she and Glenn worked so tirelessly to buy. L to R: Austin, Tom, Tom's wife Hayley, Vincent, Glenn, Tessa and Roma, with Hayley and Tom's dog Marley.

The biggest shearing they ever had on Tirlta was in 1989, when they produced 863 bales, thanks to a good season and good lambing. After the last drought production dropped to 50 bales. Ian and Merry Jackson are pictured at the height of the good times, with their three children Sarah, Matthew and Andrew.

The arrival of rain at Tirlta is cause for celebration, even if it does make driving difficult. Ian Jackson took visitor Denise Mulhlan on a tour of the property to check the tanks after an overnight storm in 1993.

American Field Service (AFS) students get a taste of station life on Tirlta, helping owner Ian Jackson castrate a calf that had been 'bull dogged', or caught by hand.

The shores of the ephemeral lake on Tirlta are a favourite spot for a swim and a beer. Merry and Ian Jackson are joined by their daughter-in-law Sara, grandsons Archie and Sam, son Matthew and English backpacker James, who was working on the station.

This sheep was killed to provide meat for Tirlta Station. Skinning the sheep after the kill was just one of the eye-opening tasks for AFS students staying with Ian Jackson in 2011.

It was all for one and one for all on the Marriott family farm when the children were young. 'This farm only survived because of the energy and input of everyone involved, from a very young age,' said mum Cath Marriott. Here Tom, Hannah and Catherine pitched in to help after shearing.

With only four years separating them, Marriott siblings Hannah, Catherine, Tom and Charlie formed close bonds growing up on Yarallah, Victoria. Chores were shared, they made their own entertainment and all four chose to study agriculture.

Cath Marriott raised four children alone and kept the family sheep farm in northern Victoria going after the death of her husband, John, at the age of forty. A committed environmentalist and animal lover, Cath is passionate about growing food sustainably and nurturing the land she farms. She's pictured with working dog Pip.

Author Deb Hunt, pictured on the South Coast of New South Wales in January 2015.

through. She would have to log on again later when the children were in bed and chase it up.

She may have had a great set-up, working from home in a job she loved, but with three children under the age of ten there was little time for relaxation.

Jo sighed. 'I would be constantly apologising for the noise if my kids lived in a city. Out here we're lucky. We've got all the space they need to run wild.'

Hamish gave up the chase and came to sit at the dining table to finish his homework. According to Jo, Hamish had outgrown the crazy energy he'd had as a toddler and he'd matured into an outgoing, sensible boy who loved footy and farming, in that order.

Annie was a quietly robust child, a dreamy artist who loved painting and swimming and who stood head and shoulders above her twin brother, Seb, even though he was older by three minutes. Her soft female energy must have been a welcome antidote to the boisterous energy of the boys. Seb was a cheeky imp, gifted with a brain that loved maths, a body that understood balance and a restless need for movement and attention. He loved motorbikes and winding up his siblings (and, all too frequently, his mother).

Annie may have had less innate ability than her twin but her greater willingness to jump in and have a go more than made up for any lack of natural talent. Jo believed Seb would succeed only when he could get over the fear of failure that held him back, a conservative approach to life he'd inherited from his mother. With an older brother who could beat him at everything and who carried no such fear, and a twin sister who was taller, stronger and more confident, it wasn't surprising that Seb played the joker. It was a role he played with mischievous glee.

Annie coughed and Jo scanned her daughter's face.

'Does your chest hurt?'

Annie shook her head. She seemed listless and I could see

Jo was worried. Jo explained what she was looking for – telltale dark lines around her daughter's eyes and a 'tracheal tug' at the base of her throat that would indicate her lungs were struggling. If she developed a temperature it would mean another trip to the emergency department in Perth.

To the casual observer Annie looked more vigorous than her brother but when pneumonia struck, as it did at least two or three times a year, she spiralled down fast. The attacks always happened at night.

'Once Annie enters the hospital system we don't see daylight for a week.'

I suspected Jo wouldn't get much rest that night.

In the last attack, heralded by a high temperature and vomiting, Jo had driven her daughter to the small local hospital in Cunderdin, eight kilometres away. They immediately put Annie on oxygen. Once that happened strict protocols came into play and they were only allowed to travel onward to Perth by ambulance.

There were no volunteer ambulance drivers on duty that night so the hospital had been forced to call the Flying Doctor, then in a moment of Pythonesque absurdity they'd had to follow protocol and find a volunteer ambulance driver who wasn't on duty who would be willing to get up in the middle of the night and don a uniform in order to drive them three kilometres out to the airstrip to meet the RFDS plane.

Cunderdin was close enough to Perth to drive there and back in a day on fast sealed roads – as Jo said, it was hardly what you call remote – but in healthcare terms, access to medical care diminished rapidly once you left the densely populated coastline.

Jo drifted in and out of sleep that night, as she must have done on many other occasions, wondering if they were about to make another emergency dash to Perth. When it happened she would sometimes call on her close friend Nat for help, but Nat had worries of her own having just been diagnosed with breast cancer.

Jo's parents lived many hours away but at least she knew she could rely on her father-in-law. At sixty-four, Malcolm still worked on the farm and lived close by. He was more than capable of looking after Seb and Hamish if necessary until Dave got back from his latest trip.

As dawn broke over the winter landscape I watched crops emerge from the mist in eerie splendour, vast paddocks of green and gold stretching towards the horizon. A crop of barley spread a green blanket over a far hill and flowering canola planted just a few metres from the house almost touched the fence line of the garden, marooning us in a sea of yellow.

Thankfully Annie's cough turned out to be the tail end of a mild cold.

The next morning's frenetic rush was no different from any other school day. Lunches were made, rucksacks packed, school books ticked, shoes retrieved and jumpers reluctantly donned before Jo hurried her three children out to the car. Seb crammed the last of his toast into his mouth then clambered onto his twin sister's lap, squashing her in an attempt to win an argument over who should sit in the middle.

'Seb, you're a smelly boy,' Jo said gravely, making her youngest son collapse in a fit of giggles.

We dropped the children at the end of the farm track to wait for the school bus that would sometimes detour up to the house if they were late (or it was early) and Jo had a quick chat with the driver about her favourite subject – the weather. They both agreed that the prospect of rain was a forlorn hope.

Over a bowl of muesli in the exquisite silence of her empty house (Jo always waited until the children were at school to enjoy breakfast and an uninterrupted coffee) we talked about how and why she had moved to Cunderdin.

'I had no intention of being a farmer and I didn't plan on marrying one either. I was duped!' she claimed.

To be fair, Dave also had no plan to move back to the town where he grew up until he received news that turned their lives upside down. In 2001 Dave was told his mother, Carmel, had been diagnosed with terminal breast cancer.

Almost immediately, Dave left Melbourne. Within weeks he'd been offered an attractive job with the Grain Pool of Western Australia and he relocated to Perth, a move that put him closer to his family and the sheep and cropping property where he'd grown up on the outskirts of Cunderdin, two hours inland. Jo was left to ponder her future.

She and Dave hadn't been living together in Melbourne but after two years their relationship had proved serious enough for them to contemplate buying a property together. Jo loved her job and she loved Melbourne, but she also loved Dave. After six months of indecision she accepted a lower-paid job in an industry that didn't interest her and moved back to Perth. Carmel died the day she arrived.

Over the next twelve months Dave's family struggled to come to terms with their loss. Dave's father was especially hard hit by his wife's death and as he mourned her passing his enthusiasm for farming began to wane. It was a solitary occupation at the best of times but without his wife by his side it was also a lonely one. Malcolm, at fifty-five, contemplated giving up farming altogether.

Dave's family hadn't got around to sorting out a succession plan. It was the last thing on their minds as they struggled to come to terms with their mother's illness and subsequent death but twelve months on, with Malcolm considering early retirement, the issue of succession could no longer be ignored.

From the moment they could walk and talk, Jo and her siblings had had it drummed into them that only someone willing to work on the family farm could expect to inherit it. 'Whoever comes back onto the farm will inherit the vast majority of it,' her father would repeat, time and time again. His insistence was designed to pre-empt any arguments about succession later on. 'I love this land and I want to be taken off this farm in a box but no amount of dirt is worth destroying family relationships for,' he would say.

Jo had no desire to farm so she accepted that she would only ever inherit a small part of it. Her brother felt likewise. Surprisingly, it was Jo's academically minded sister, Kerryn, who was married to a half-Singaporean man raised in Perth and had no background in farming, who turned out to be the one who wanted to farm.

'That meant Kerryn and Frank took responsibility for my parents' retirement, and that was a far bigger responsibility than any financial gain.'

In Malcolm's family, it had been Carmel's dying wish that the farm be split equally between her three children. It was an admirable wish but granting it would have meant selling a farm that had been in the family for generations. Times had changed since Malcolm's parents were able to afford to put five children through boarding school on the proceeds of a farm far smaller than its current acreage. Profits had fallen, margins had been squeezed and, in spite of its increased land size, the farm would not have been viable if it were split in three.

After his mother's death, Dave and Jo bought a house in Perth and got married. Their hectic urban lifestyle continued, if anything ratcheting up a notch. Dave had studied agribusiness at university (the only one of the three children to do so) and he loved the outdoors. A desk job with the Grain Pool that trapped him behind a computer screen was far from ideal, especially when he was sometimes still at his desk at midnight, trading grain on the Chicago Board of Trade. Dave released his stress the only way he knew how – by playing as hard as he worked. Weekends were spent participating in extreme sports, barefoot skiing when it was barely two degrees, and partying hard.

Jo had no doubt that Dave was heading for an early heart attack if he didn't get out of the job he was in, so when he mentioned he'd been contemplating a move into farming she felt relieved. Jo's mother had also been diagnosed with cancer by that stage, a sobering reminder of the fragility of life. Jo wasn't thrilled at the idea of leaving Perth – she had friends and a lifestyle she enjoyed – but she didn't love her job and she'd always planned to stop work when they had children anyway.

Dave's brother was a podiatrist and his sister worked as a counsellor, and since neither of them had any intention of changing profession, Dave was the obvious candidate to take over the farm. The family nutted out a succession plan, helped by an impartial consultant who guided them through the tricky process of reaching an agreement. Raising the money to buy a two-thirds equity share was beyond Jo and Dave's means so they came up with a complex business structure that would allow Dave and Jo to inherit the farm over a number of years. It was agreed that Malcolm would have a place to live on the farm for as long as he wanted, and Dave and Jo agreed to lease certain paddocks from him to guarantee he'd have a steady cash flow in retirement. As in Jo's family, with inheritance of the farm came responsibility for Malcolm's retirement.

By the time the deal was signed, Jo was two months pregnant.

In early 2004 she and Dave resigned from their jobs in Perth. For better or worse, they were embarking on a new life in the country.

It was better for Dave right from the start. He took to farming like the proverbial duck to water, putting into practice everything he'd learnt at uni and during his years with the Grain Pool. He was back on the farm where he'd grown up, his best mate lived in the next town along and he was working alongside his dad, a man he loved and respected. No longer trapped in a stressful office, Dave suddenly had wide stretches of land to farm, an impressive array of machinery and the prospect of being his own boss.

For Jo, who was now three months pregnant with their first child, there was no better, there was only worse.

'I hated it.'

Jo's first impressions of life on the farm weren't helped by the weather. Throughout their first ten days, daytime temperatures didn't once drop below forty; day after day the sun blazed down, scorching the already parched earth and trapping the stifling heat in the flimsy fibro house they'd moved into.

The transportable (the one I'd spotted on arrival) had been hastily erected in 1969 following a massive earthquake that had devastated much of the central wheatbelt area the year before, including the beautiful old mud brick house that Dave's grand-parents had once lived in. Dave's mum, Carmel, had been so embarrassed by the quality of the replacement that she'd been reluctant to entertain visitors. Forty-four years after it had first been installed, Dave and Jo moved into the ramshackle house in which he'd spent much of his childhood.

The house had been vacant for some time. Far worse than the flimsy building, with its lack of air conditioning, collapsed ceiling and asbestos fibro walls, was the infestation of rats and mice that greeted Jo on her arrival. She spent the first few days in her new home on her hands and knees, sweeping up piles of rat drop-pings and cleaning out kitchen cupboards that stank of urine. Wearing rubber gloves she would reach into stinking cupboards, stop occasionally to stand up and vomit into the sink, then go back to scrubbing out the cupboards.

I remember the shock of seeing plagues of crickets in Broken Hill, followed by plagues of mice. The whole town was affected and we simply had to wait it out until they moved on. Rats and mice were nothing unusual on a farm homestead, but suffering from chronic morning sickness as she was, Jo found the clean-up challenging. After two years in the heart of happening Melbourne – 'the best years of my life' – followed by three years in a beautiful suburb of Perth, Cunderdin must have been a shock.

The barren landscape around their house was scorched and ugly, devoid of crops in the height of summer. Memories of the exquisite garden her mother had cultivated in the more temperate climate south of Perth were wiped out in the face of extreme heat, dirt, snakes, spiders, lizards and clouds of mosquitoes. When she didn't have her head in the sink due to morning sickness, Jo was trapped indoors because of a severe allergic reaction to mosquitoes. Her skin ballooned up whenever she was bitten.

There were no antenatal groups in Cunderdin and no prospect of a mother's group to join after her baby was born either.

'Initially I really struggled here. I hated the heat, the landscape, the dirt, the house and I missed my friends and family.'

Jo was desperate to escape. Her only salvation was the occasional trip to Perth for an ultrasound.

'I hung off those trips. I spent hours working out how to get back to live there.'

Jo spent her first eighteen months in Cunderdin planning ways to escape. Her escape plans sometimes involved her husband Dave.

'And sometimes they didn't!'

While Jo struggled, Dave went from strength to strength. Having left the stress of corporate life behind, he was a changed man: happy, outgoing and more relaxed than Jo had seen him in years.

Dave's laid-back father, Malcolm – a man of quiet strength and gentlemanly manners – had worked alongside his own father for the best part of twenty years so he understood the journey his son was on. Graciously acknowledging that Dave's university degree, coupled with the progressive farmers he'd dealt with in his grain-buying days, had equipped him with ideas worth listening to, Malcolm agreed to relinquish the stud ram business he'd built up as insurance against crop failure. Dave argued that the time-consuming chores of moving sheep from one paddock to another, checking water troughs, feeding, drenching and fence-mending weren't worth the financial return.

'It's a lot of effort for not much money,' he told his father.

'That's true,' said Malcolm, smiling inwardly. It was also true that his son hated sheep.

Sheep were the only farming Jo had ever known. Her parents were third-generation sheep farmers and Jo remembered sitting on the floor as a child, watching her grandmother tease out threads of wool on a spinning wheel. The bed socks her grandmother gave her, knitted from wool she had spun and dyed by hand, were still in use. For all that, she agreed with Dave's decision. His argument for a greater focus on cropping was based on lifestyle, not nostalgia.

'By getting rid of the sheep we've got more freedom,' he argued. 'We can work hard in the peak months and take time off when it's quiet. We'll be able to go away in January and February.'

Jo couldn't argue with that. The first two months of the year were ugly, no other word for it. There was nothing but endless heat, barren earth and not a blade of green to be seen. She would happily have been anywhere but Cunderdin in January and February. Or the rest of the year come to that.

How did Dave react to how she was feeling, I wondered?

'Dave knew I was having a hard time but we'd agreed that this was what we wanted to do, this was where we wanted to raise our children, so what could he do? He just had to wait and hope I'd settle.'

It was a long wait. At first, Jo barely left the house, doing what she could to make the transportable more comfortable. The rats never left and on one occasion Hamish accidentally swallowed rat poison. It was one of many times when Jo rang the poisons helpline.

'I had them on speed dial for a while.'

Eventually she was forced to accept that they would have to live with the odd rat in the rafters.

There were two bright spots on the horizon. The first was that her mother's cancer was in remission, and the second was play-group. Once Hamish was six months old Jo took him every Friday morning to a small, rundown building in the centre of Cunderdin.

Less than a year later, she fell pregnant again, this time with twins.

Jo's second pregnancy couldn't have come at a worse time. Dave had been awarded a Nuffield scholarship, which involved heading overseas for several months to study farming practices around the world. Jo sweltered alone through another unbearably hot summer, suffering what felt like a double dose of morning sickness, and this time she had Hamish to look after as well.

The Nuffield scholarship entailed two two-month stints over-seas and, as luck would have it (Dave's luck, not Jo's), the second stint coincided with the last trimester of Jo's pregnancy. Carrying twins took its toll on her slender frame. When Jo reached the point where she could no longer reach into the cot to pick up Hamish, she was forced to move back to her parents' farm – a three-hour drive from Cunderdin – until Dave returned.

'Dave got back about three weeks before I gave birth and I pretty much gave up walking from then on.'

Twins Sebastian and Annie were born in Perth in October 2006 and Jo spent the next eight days in hospital.

She went home to the transportable in Cunderdin with newborn twins and a small toddler to look after. Less than two weeks later Dave started harvesting, and Jo effectively said goodbye to him again. Harvesting in the central wheatbelt of Western Australia was a twenty-four hour operation.

'Once it starts you don't see the men at all.'

We took a break from our interview to drive into Cunderdin and pick up supplies. Yet again, Jo was full of fascinating facts. She pointed out the pumping station on the main road through Cunderdin. 'That pipeline you saw was designed by a man called O'Connor in the late 1800s to take water from Mundaring Weir all the way up to Kalgoorlie. When they turned on the tap in Kalgoorlie it didn't work, so he rode into the water at Freo and drowned himself. The water arrived a couple of days later.'

That was the pipeline I'd followed on the drive from Perth.

'You might want to research my facts though.'

Jo was a true journalist, the way she encouraged such corroboration. Much of what she told me turned out to be correct. O'Connor was a brilliant engineer, driven to suicide in 1902 by parliamentary criticism and vicious press attacks on his competence. He rode his horse into the water then shot himself.

Jo confessed over lunch that the only thing she hated more than snakes (and she really, really hated those) was drawing attention to herself, which made it all the more stunning that she had been willing to front a media campaign in 2009 that put the spotlight squarely on her and the town.

'I did it because I was livid. I was fighting for women like me who had no support and I was furious that politicians and bureaucrats in Canberra could overturn a policy that had such a massive impact on our lives.'

The policy she was talking about was connected to childcare rebates. For two years, with three children under the age of two, Jo worked behind the scenes to establish Cunderdin's first child-care centre. It was a remarkable investment of time and energy from someone who 'hated' Cunderdin when she first arrived and had felt isolated in the community.

I soon learnt that when Jo Fulwood latched onto a cause she was like a kelpie with a kangaroo bone – she wouldn't let go until the bone had been licked clean.

The story of how she, and others, made it happen was incredible.

*

Jo clipped three-year-old Hamish into his booster seat and went back inside, swatting away the swarm of mosquitoes that always seemed to hover around the entrance to the house. The rattling air cooler they'd finally had fitted was struggling to cope with the summer heat and she felt exhausted. Jo's energy levels plummeted to such depths in summer that she often wondered if she was suffering from some kind of chronic illness. Come winter, when frost sometimes hung in the air and it was colder inside than outside, she would regain her energy and realise she had been suffering heat exhaustion.

She loaded herself up with the double buggy, nappy bags, toys for Hamish, hats, water, handbag and sunscreen then trudged back out to the car, keeping a wary eye out for snakes as she picked her way through the capeweed that straggled along the path. Eggs were no longer an attraction since a fox had broken into the yard and savaged the chooks but snakes were an ever-present threat in summer.

She piled up the car, ducked back in for the twins and settled them one by one into their car seats. In another hour or so one of them would need to be fed. She might just have enough time to get Hamish to playgroup and say hello to some of the other mums before Seb or Annie would wake up grizzling.

Jo had always said that she wanted to be a stay-at-home mum while the children were little and she'd held fast to that position, but she hadn't expected to have three children under the age of two and a half. In Perth it would have been easy, she would have had many options in long day care, preschool, family day care, early learning, Montessori, private nannies, babysitters or nurseries. In Cunderdin, apart from a single weekly playgroup on a Friday morning, there were no childcare facilities. None.

She drove down the farm track, turned right at the entrance and left onto the side road, and pulled up at the T-junction, waiting for a break in the thundering convoy of trucks carrying mining equipment to Kalgoorlie, before turning right onto the fast highway. Two kilometres later she pulled up outside play-group on the main street of Cunderdin.

Hamish stood beside the car, on the shady side, as she unloaded the double buggy, settled the twins and grabbed the bags. The dingy room that housed playgroup was brightened by posters and children's drawings plastered across the walls. Jo smiled at the mums she recognised.

When she spotted Nat her spirits lifted. Nat was a few years younger than Jo and she had three young children. A prag-matic, generous woman with a blunt sense of humour, Nat lived twenty minutes north of Cunderdin. By way of connection, Nat's husband was Dave's second cousin, and she and Jo had formed a strong bond when they discovered they both had young children with limited support.

Jo would never forget the day when, just before Hamish was born, Nat turned up on her doorstep with a bag of nappies.

'Charlotte doesn't need these anymore,' she'd said. 'I thought you might be able to use them.'

There had been the odd phone call and the occasional email after Hamish arrived and then, when Jo was heavily pregnant with the twins and had been forced to leave Cunderdin to spend time with her parents, Nat had stayed in touch.

Jo would never dream of admitting it to her in person (Nat wasn't the sort of 'lovey-dovey' person who would take kindly to an overt show of emotion) but Nat had made Jo's life in Cunderdin bearable. It was as simple and as stark as that.

'Hi Nat, how's life?'

'Isabelle is teething, and Charlotte's got a cold. Situation normal, I'd say.'

Jo laughed. Nat could always be relied on to see the funny side of life.

'Did you hear kindy's moving?' Nat said. 'They're amalgamating into the main school.'

'Any idea what's happening to the kindy building?'

'I think a youth group wants it.'

'Unless someone opens a childcare centre.'

'Hah, in your dreams!'

Nat wasn't to know but such a possibility *was* in Jo's dreams, and it had been for a while. She had been shocked to discover what little support there was for new mums in Cunderdin. It was a shire town that served a population of 1300. The local supermarket was well stocked and there was a thriving post office, a local hospital and a full-time doctor's surgery. So where was the support for new mums? Where was the childcare?

It wasn't hard to pick up on an undercurrent of tension running through Cunderdin. The fractured sense of community was partly explained by the presence of the Exclusive Brethren, who led a distinctive and separate way of life.

The Brethren had been well established in Western Australia since the 1930s. From the late 1980s Brethren communities migrated from Perth to settle in rural areas such as Dalwallinu and Cunderdin, where it was felt they could better distance themselves from what they considered to be the world's evils. Brethren women were easily identified by the scarves most of them wore, and Jo soon learnt that their children did not socialise with others from outside the faith. When it came time for morning break at the local pre-primary, Brethren children were segregated, given their own plates of fruit and made to sit apart. From grade three they attended a separate, Brethren-run school.

Jo was convinced she would have got on well with some of the Brethren mums if only they'd been allowed to socialise, but she'd been forced to accept it would never happen. That still didn't explain why no one had thought to set up a childcare centre. Jo reasoned there had to be plenty of women who went to work and needed childcare, as well as other stay-at-home mums like herself who just needed a break sometimes. How did they cope?

'Would it work, do you think?' Jo asked Nat.

'What?'

'A childcare centre?'

'Are you serious?'

'Why not? If the building exists, we're halfway there already.'

Jo went home with the glimmer of an idea. If she could drum up community support, get the necessary approvals, raise enough money, apply for grants and do the work involved in refurbishing the building then a childcare centre for Cunderdin might just be possible.

Even though it was a long list of ifs, Jo felt a surprising surge of enthusiasm. She'd had plenty of experience running campaigns and making things happen in the world of PR and marketing, and now that the twins were a year old she was itching to get stuck into a project again. Far from feeling overwhelmed, she felt energised at the prospect of getting a childcare centre up and running. Dave was a member of Lions, he could get a working bee together and Nat would help, she was sure of it. There had to be plenty of other women in Cunderdin who would benefit from a childcare centre – they might be willing to lend a hand as well. She resolved to talk to Dave about it when he got back from another of his overseas Nuffield tours.

She turned her attention to the architectural plans for their new house. They'd had an agreement with Malcolm that he would move out of his brick house within five years and they would move in. The five-year deadline was fast approaching and she and Dave were reluctant to enforce the agreement. Malcolm had found a new lease of life farming alongside his son and he was a hands-on, loving grandfather. He had a special rapport with Jo's children and he'd even got married again, to a nurse, Jovita. Turfing him out of his home now would only push him to move to Perth, and no one wanted that to happen.

The answer was for Dave and Jo to build. After several years renting out their Perth house they'd finally taken the plunge and sold it, turning a tidy profit thanks to the mining boom. That profit would just about cover the cost of building a new home.

Jo felt a welcome return of the energy and focus that had propelled her through a successful career in PR.

Yes, the man on the other end of the phone said, the old building will be available when the kindergarten shifts into the main school, and no, the council hadn't made any decisions yet on what to do with it.

'The local youth committee is interested,' he added.

Jo moved swiftly. First she went to see the local community development officer, an energetic woman with a reputation for getting things done.

'I think it's a great idea,' she said. 'I can write a business case in support of the childcare centre, as long as you can prove there's enough support in the community.'

Jo called a meeting.

What will it cost? Who's going to pay for it? When will it be ready? The questions came thick and fast and there seemed to be overwhelming support amongst the women who turned up to the meeting, although only a few were willing to put up their hands to help. Nat came good, as Jo knew she would, and several others offered to serve on a small committee. It was enough for Jo to believe the project might get off the ground.

Not everyone thought it would succeed, or even deserved to. Jo met one woman in the local supermarket who eyed her suspiciously and said, 'Why do you need childcare? I never had it in my day and I survived.' The stranger pushed her trolley away before Jo could think of a suitable rejoinder.

Over the next few months Jo got her head around the myriad regulations and endless red tape that governed the operation of officially sanctioned childcare centres. From the outset it was important for the centre to be approved, because only then could parents apply for a reduction in fees. That was vitally important for a town like Cunderdin that, like so many others across the wheatbelt community, was in a low socio-economic area. If it was an approved centre the proposed childcare fees of $60 a day could be cut in half, and that would make a big difference to most parents. What's more, Jo and the committee could also apply for financial assistance to run it.

Under pressure from the community, the shire agreed to hand

over the building at a peppercorn rent for use as a childcare centre.

'The youth committee wasn't too happy, but you can't please everyone,' said Jo when she told Nat the good news.

Stringent new legal requirements meant the old building had to be gutted. Old ceiling fans that were now too low had to be ripped out, the perimeter fence had to be replaced because it wasn't high enough, they had to dump the pine playground equipment that had rotted, fit new gates that locked . . . the list went on and on. Dave and his fellow Lions members helped organise a working bee and volunteers emerged to offer their time and labour. When Lotterywest came good with a $75,000 grant to help fund the cost of refurbishment works the project gathered momentum.

For all the physical progress everyone could see, there were hours of administrative tasks behind the scenes. It was a long, arduous process, especially with three children to look after. Jo was buoyed by the occasional comment from parents in the supermarket. 'I hear that childcare centre's going well.' 'When are you opening?' 'Can I put my kid down for a place?'

Jo didn't intend the centre to be purely for women who wanted to go back to work. It was also for people who simply needed a break from their kids occasionally. She was determined that women like herself, who had no local support network, should have access to quality childcare.

It was clear that the community of Cunderdin wasn't large enough to support a centre that would open five days a week so they set a target of three full days, operating from eight to five. There were plenty of examples of other centres in rural areas that operated similar hours, and they all met the criteria for the vital Child Care Benefit.

With Nat on board to share the burden, and a team of committed volunteers, the project inched towards an opening date. One of the last hurdles they faced before any staff could be recruited was a shortfall of around $35,000 to help solve some of the regulatory issues. Jo approached the local council and lodged an application for funding.

She and Nat were called to a council meeting to argue their case, meeting local councillors from farming and non-farming communities. Jo couldn't help noticing that there was only one woman among them.

She needn't have worried. Everyone could see what this team of volunteers had managed to achieve and there was general consensus that their application should be approved.

'Congratulations on all you've achieved, and good luck with the opening,' said their lone female councillor as Jo and Nat left the meeting.

By early September 2009 they were just days away from opening. Ads for staff had appeared in the local paper, the final coat of paint had been applied, furniture had been delivered, turf had been laid and the brand-new building was almost ready for its first occupants. Jo spent most of her day on the phone, project-managing the building of her own house then chasing supplies for the childcare centre, checking permits and arranging interviews for new staff. It was only when she'd given the children lunch and put them down for an afternoon nap that she found time to open the post she'd collected from Cunderdin earlier that morning.

She slit open an official-looking envelope from the department of the Federal Minister for Childcare in Canberra and scanned the contents. Her stomach lurched as she read the letter, re-reading it twice to make sure there was no mistake.

'Legislation has been revoked,' the letter stated. 'Childcare centres open for less than five days a week will no longer be approved for the Child Care Benefit, effective immediately.'

Jo knew exactly what that meant. Parents putting their children into childcare centres open less than five days a week could no longer expect to receive any reduction in fees, no matter what their financial circumstances. And such centres would no longer qualify for any financial assistance. The changes would have devastating consequences for Cunderdin. The future of their new childcare centre was in jeopardy before it had even opened.

This unexpected announcement sent Jo into a tailspin. She rang the number at the bottom of the letter and spoke to an official whose belligerent tone suggested Jo wasn't the first to call and complain.

'The decision is made and nothing can be done about it. This call is a waste of time,' the official said, bluntly.

Jo matched the woman's aggressive tone. 'That is a ludicrous decision,' she shouted. 'What do any of you know about childcare in rural areas and our requirements?'

She slammed the phone down and rang her politically astute parents.

Jo's mother was well aware of the work her daughter had put into getting a childcare centre established and she listened with growing concern as Jo described the situation.

'Ring the federal minister,' Bev said. 'Tell her what the issues are.'

The closest Jo could get to the federal minister was an aide. Yet again she was fobbed off by someone who showed little concern for the issues facing a small community in the central wheatbelt of Western Australia.

Jo put the phone down and rang Nat to break the bad news.

'But the centre can't afford to operate five days a week!'

'I know.'

'All of our costings are based on operating three days a week.'

'I know! And if we let them get away with this the centre won't be viable. Those bloody politicians could force it to close,' Jo said, still smarting from her earlier conversations.

'What can we do?'

'We can wage a war.'

Following her mother's advice, Jo got political. She enlisted the help of Mia Davies, MP for the Western Australian agricultural region, and Brendon Grylls, Minister for Regional Development in Western Australia. Together they lobbied the federal minister for childcare, who turned a deaf ear to their pleas. When it was obvious the minister wasn't listening, Jo embarked on a PR campaign that was as committed, focused and passionate as any she had run in her professional life, only this time it was personal.

Furious at the lack of response from those in authority, she wrote blistering letters, issued carefully worded press releases that were damning in their conclusions, and conducted interviews on local radio and on television. Her message was simple: the federal government was forcing childcare centres in rural areas to close.

'It's unthinkable that the government is not willing to consider some flexibility in the legislation to ensure the long-term survival of these essential services in country areas,' she told *York & Districts Community Matters* in June 2010.

In a letter to the prime minister, she wrote:

We are passionate about the development of our community and committed to the long-term survival of our childcare service. Our community, and families in our community, will be devastated if our childcare service is forced to close . . . Unfortunately, this is the most likely outcome if legislative change does not occur . . . We hope you and your government . . . will immediately move to change this draconian legislation.

Jo unleashed a barrage of criticism of government policy.

'I've got nothing to lose,' she told Dave when he expressed a modicum of concern. 'I'm sick of bureaucrats in Canberra thinking they can ignore us.'

Dave knew that his wife wasn't a quitter. Once Jo got attached to an issue she wouldn't let go until she'd won it. The interviews continued and the flurry of emails and online messages gathered momentum. It must have been clear to anyone listening in Canberra that Jo Fulwood wasn't going to shut up anytime soon.

Success came in the form of a phone call from the federal minister's office.

'We are not changing the legislation, you must understand that,' said the faceless bureaucrat on the other end of the line. 'However, the minister is prepared to offer the childcare centre in Cunderdin an exemption to the new policy.'

'Thank you,' said Jo. 'I look forward to receiving that in writing.'

By the time Brendon Grylls and Mia Davies attended the official opening, on a bright spring morning on 21 September 2009, Jo's twins were less than six months away from starting kindergarten, and recurrent pneumonia had prevented Annie from ever being allowed to attend childcare on a regular basis.

Ironically, Jo barely got to use the centre she had fought so hard to establish.

*

It's doubtful if any of the parents who now use the childcare centre in Cunderdin realise the work Jo, Nat and the other volunteers put into making it happen.

'And that's as it should be,' said Jo. 'We won the fight, that's what mattered.'

Jo had told me the full story over several glasses of wine the night before, often downplaying her role in the process to praise others, especially Nat.

'I hate trumpeting that I was the person who set it up. We had massive community support.'

That was no doubt true, but every cause needs a champion and Jo Fulwood championed the childcare centre in Cunderdin. And as Brendon Grylls said at the opening, 'The campaigners left an important legacy.'

We were sitting in Jo's open-sided sunroom, shielded by mosquito netting on all sides, enjoying a late breakfast without fear of interruption now that the children were at school, and Jo paused in her contemplation of the garden.

'I'm passionate about the importance of agriculture and about arming women in regional areas with the skills and opportunities they need to develop their potential,' she said, unexpectedly revealing the depth of her feelings.

Her keen interest in the role of women in rural areas, ignited initially by the childcare campaign, led to a board position on RRR – the Rural, Remote and Regional Women's Network of Western Australia – and that led to her current job as a journalist with the *Countryman*. It was obviously a job she loved, and not just because she could work from home.

The interviews she conducted sometimes left a lasting impression. A young, fifth-generation farmer responded to one question about political influence with a quote that lingered.

'We don't have any,' he said, bluntly. 'In third world countries, when there's a massive drought it translates into a lack of food for people who live in cities. In Australia, when there's a crisis in agriculture it has no impact on the urban environment. We are of no significance to most politicians or to the majority of consumers.'

We are of no significance. It was a damning assessment and probably true. How many of us think about farmers when we reach for food on a supermarket shelf?

Some of the farmers Jo had interviewed were hollowed-out men defeated by a rapidly changing climate that only got drier the further east they went, with frost a growing problem.

'Dave's grandmother can never remember having frost on such a regular basis, and we've had frost every year since we got here.'

Jo may have been a passionate advocate for greater recognition of farming and farmers but she also had a sense of humour. We nipped out to film a short clip of her walking through a field of canola – something the board of RRR had asked her to submit – and I jokingly suggested an alternative version.

Jo giggled. 'Like *Gone with the Wind*?'

In a flash she was skipping down the farm track, scarf floating in the breeze, doing a slow-mo impression of Vivien Leigh (or Dorothy!).

Later that afternoon Jo pulled out a batch of scones from the oven and we left them cooling on the benchtop while we picked up the children from school, getting back just as Dave returned from his latest trip to Perth. The children rushed to greet him and there was a scrabble for his attention until they noticed there were scones to be had.

'I was raised by a farmer's wife who preserved fruit, baked cakes, bottled jams and made all her children's clothes,' said Jo, placing a pot of homemade jam in the middle of the table. 'If I send mine off to school without a homemade cake or a slice I feel like I've robbed them of their childhood.' She was joking (I hope).

Malcolm dropped by to join the rambunctious tea and his quiet presence had a calming effect on the children.

It was an ideal opportunity to ask Dave and his father how they both felt about farming. The big picture for Dave was sustainability.

'I want each generation to have a better standard of living than the last. If my kids want to farm I want them to have that option.'

The first generation of Fulwoods to farm at Yarrandale lived in a mud hut with hessian bags for walls. Dave's grandfather was the next generation. His wife, Syvlie, was one of thirteen, and went on to have five children of her own. Quite apart from providing three cooked meals a day, plus morning and afternoon tea, her life revolved around cleaning, washing, sewing and bringing up the children.

Dave said he will insist the children study something else first before making any decision about following the family tradition.

'Farming is so complex and diverse today. Apart from being a farmer, you have to be a mechanic, a bookkeeper and a trader, and there are big risks involved. It will be good for them to have a back-up plan.'

I couldn't help thinking that no parent would dream of suggesting to any budding lawyer, doctor, pilot or policeman that they have a back-up plan. Farming was a risky business.

Dave's work with Nuffield put him in the unique position of being able to explore farming operations in other countries, and since he took over as president in 2012 he had regularly led tour groups through Europe and Russia.

'If there was somewhere else in the world where they were making more money and had less risk, we'd move there.'

So far, despite the challenges of dryland farming in an area officially classed as desert, Dave hasn't found anywhere with the same benefits that outweigh the risks, pointing to the political and economic stability farmers in Australia often took for granted.

Malcolm was glad he was no longer in charge. Not long after Dave and Jo moved into their new house the family threw a party, in September 2010, to mark the farm's centenary. It wasn't an ideal time to celebrate – the house had only just been finished, the garden was a dust bowl and the crop was pitiful – but everyone turned up to mark the occasion, including Dave's grandparents. I was beginning to understand that in farming communities they were all in it together, all suffering through the same conditions.

Not long after those celebrations, Malcolm decided Dave was capable of taking over. From that day on, he stopped going into the office.

'It was a relief. I still watch the forecast and I still wish for rain, and I probably always will, but that awful anxiety has gone.'

Now it was Dave (and Jo) who lay awake at night worrying about the threat of frost, and they were eager to see if mould board ploughing might be the answer. One of their neighbours had only managed to plough half his paddock using the technique, and when his crops suffered a snap frost there was a line down the middle, black-frosted on one side, frost-free on the other.

They were hoping for a good harvest too. It would be another few weeks yet before the threat of frost passed, so in the meantime they watched, and waited, hoping for a few more millimetres of precious rain.

Dave admitted that harvest time was special, even if he did have to put in twelve- to eighteen-hour days. Hamish loved to join his father on the John Deere 8420 that pulled the chaser bin, or the New Holland combine harvester. Watching the header unload while it was still harvesting, auto-set to travel at precisely the same speed as the tractor and the chaser, was a magical sight for everyone, especially when it happened under moonlight. Seb suffered from hay fever so he normally stayed well away and Annie hadn't shown that much interest yet, and with so much heavy machinery involved the children were under strict instructions not to approach the working areas unless an adult accompanied them.

'A farm can be a dangerous place,' said Dave. 'It's like a factory without walls.'

The children were growing up fast. In another three years Hamish would be at boarding school and Dave was beginning to realise how much he would miss the kids when they left home.

'We've got to make the most of them now, before they decide we're not cool anymore,' he laughed.

Dave recently floated the idea that they could look at moving back to Perth when Hamish reached high-school age.

'I could drive back here to farm during the week,' he'd said, thinking Jo would welcome the idea. She was horrified. Her instant reaction was, 'No way! I'm not moving. I love my job, I love my home and I love this community.'

It was a huge turnaround for someone who spent her first eighteen months in Cunderdin plotting ways to escape.

Before I left, Jo took me on a tour of her surprisingly well-stocked garden.

The family had moved into their new house in 2009, a year before Cunderdin experienced its second worst drought on record (a drought that lost them a shocking half a million dollars) so there'd been no money left for such non-essentials as plants. Valuing beauty as she did, Jo had been determined to create a garden and the only way was to start from scratch, using cuttings and shoots.

Friends responded to her call for help by donating trees, roses and plants. She scavenged a bag of lawn cuttings, hired a roller and laid the shreds of turf herself, transforming the once-empty paddock into a flourishing sweep of lawn that must have been a welcome sight on sizzling summer days.

Manchurian pear trees ran along the fence line and dozens of David Austin roses clustered beside the veranda. It wasn't as planned as Jo would have liked it to be, so some of the results were surprising. Drought-tolerant rosemary, lavender and native hibiscus flourished alongside more tender camellia bushes. I even spotted a hollyhock in there somewhere – an unlikely survivor in the near-desert climate of the central wheatbelt area.

Whatever Jo was given went in, including poppies, pelargonium, citrus trees, frangipani, boronia, kangaroo paws, convolvulus and daphne (a drought-tolerant variety her mum found).

It was a beautiful sight. Roses, which Jo loved, were tucked into every available space, promising a show of magnificent blooms when the country beyond shrivelled in the height of summer.

'When you live in the desert, having a little oasis is not too much to ask.'

Jo had confessed earlier in the day that the one thing she longed for was a pool, and from what she'd told me about conditions in summer I could see why. She knew exactly where she would put it too, stretching out across the lawn so it would be visible from the outdoor room.

If they got enough rain that year, and if they didn't get frost, and if the harvest was a good one – all those ifs again – she and Dave had calculated that there might just be enough money to invest in that desperately needed swimming pool.

She nipped into her office to check the weather forecast.

Michelle Reay and Jon Cobb

'Durham Downs'
350 kilometres west of Quilpie, southeast Queensland

I had already spoken to Michelle several times on the phone before we met in person, and long before the idea for this book was conceived. Back in 2011, when I was working in communications for the Royal Flying Doctor Service, Michelle's youngest son, Will, was involved in a serious accident. The RFDS responded and I wrote a brief story about what happened for the quarterly newsletter.

I was delighted to bump into Will and his mum at the Innamincka Races in a remote part of northeast South Australia several months later. I had flown there with the RFDS, who were providing medical assistance for the event, and I watched Will and his brother pick up several rosettes in the gymkhana. I couldn't help thinking how incongruous it was for a young Englishwoman from the Midlands, whose accent sounded unchanged after years in Australia, to be living and working on one of the largest and most remote Kidman cattle stations.

Knowing that Michelle and Jon both worked there, and that they had four young sons, they seemed an obvious choice to interview for this book.

Getting Michelle to agree to be interviewed was far easier than getting to Durham Downs to see her. Head west from Durham Downs and the next stop (beyond Innamincka) is Lake Eyre. The nearest commercial airport, at Broken Hill, is a nine- or ten-hour drive from Durham, up through Sturt National Park, across the border into Queensland and on towards Nockatunga in the Channel Country. Port Augusta is a seventeen-hour drive away. The mail plane was a brief option until I discovered they only stopped at Durham once a week, and there was no way I could inflict myself on Michelle for the length of time it would take for the plane to complete its circuit and pick me up again. I was stumped until the RFDS and Santos came to the rescue.

The Santos oil and gas exploration company has several plants in the Strzelecki Desert and the plant at Ballera was less than an hour's drive from Durham Downs. The RFDS held regular health clinics there, providing emergency services for Santos staff and the surrounding graziers, and it was where Michelle took the children anytime they got sick.

RFDS nurse practitioner Chris Belshaw knew Michelle and her family well. Thanks to him, I was able to join an RFDS clinic flight from Broken Hill to Moomba then on to Ballera in the Cooper Basin, followed by a Santos shuttle flight from Ballera back to Adelaide.

Durham Downs was certainly isolated. Beyond the homestead, the working sheds, the guest quarters and the schoolroom, there was no sign of human intervention. There were no telegraph lines or electricity pylons, no fences (that I could see) or culti-vated land, just the wild beauty of the Australian Outback. In that part of southwest Queensland, summer temperatures can creep towards a sizzling fifty degrees.

The challenges facing any parent working on the land in a remote area were exacerbated for Jon and Michelle because, unlike multi-generational graziers and pastoralists whose children often live on adjoining properties, Jon and Michelle were thousands of kilometres away from the support of their respective families.

It was that isolation and freedom that first attracted Michelle when she was a young English backpacker fresh out of university on the classic twelve-month stint down under.

Jewel in the crown

The Channel Country of southwest Queensland is almost 60,000 square miles of flat plains fed by the Cooper, Diamantina and Georgina rivers, which frequently spill over when heavy monsoon rains fall further north. In her 1987 book, *Kidman: The Forgotten King*, Jill Bowen described it as a 'superb natural irrigation system'.

Sidney Kidman bought Durham Downs in 1909 and it gave him a vast stretch of rich cattle-grazing country in South West Queensland, spanning 3500 square miles. He paid $100,000 (a cool $20 million in today's terms) for three properties – Berawinnia Downs, Tilbaroo and Durham Downs – and the largest and most important by far was Durham Downs.

The story of inland Australia's punishing cycle of drought and flooding rain was told in the statistics Bowen quoted: twelve years before Kidman bought it, there were 27,500 cattle and 7500 sheep on Durham Downs. By the time Kidman acquired the property it could muster only 7000 cattle. By August 1913 Durham Downs was again carrying 25,000 cattle, yet in the subsequent three years of drought thousands of those cattle died of starvation.

'A place of abject misery,' wrote Bowen, referring to Durham Downs when it was locked in drought. Anthony Kidman described it as, 'a moving sand hill for thousands of miles'.

I must admit the prospect of driving to Durham Downs filled me with dread. The seven-hour drive to Gilberton had pushed me to my limit (when I lived in England I used to pull off the motorway for a nap halfway through the two-hour drive between London and Bristol) so I knew the nine or more hours it would have taken me to drive from Broken Hill to Durham Downs would have involved an overnight sleep in the car somewhere – on the way there and on the way back. I was mightily relieved when Santos and the RFDS found a solution to the problem of how to get there.

I had no plans to visit any part of the impressive oil and gas plant that Santos operated in the middle of the desert but strict health and safety procedures demanded that I complete an online training program before my trip could be authorised. It was far more rigorous than I'd anticipated and meant sweating at the computer for over an hour; failure would have scuppered any chance of interviewing Michelle. All I recall now was a large red STOP button and the MD's stern face appearing on screen. *I authorise you, yes you, to shut this plant down in the event of an emergency.* (Me? Are you sure? I once set fire to a pair of curtains while doing the vacuuming – improbable but true – and the mess I made clearing up was far worse than the initial fire.)

Michelle was waiting at Ballera with Will, now four, who looked very different from the fun-loving boy I'd met at Innamincka Races two years earlier. The only lasting effect from his accident had been a long pale scar running under his blond fringe. But I remembered Will as a boisterous toddler full of energy, proudly displaying the yellow sash he'd won in a gymkhana event. Now he looked pale and listless, sitting in the small clinic's waiting room, clutching an uneaten biscuit.

'Poor Will's not feeling too good. His tonsils are inflamed,' Michelle said.

Knowing she would be driving in to pick me up on the day an RFDS doctor would be at the clinic – which was staffed by nurses the rest of the time – Michelle had waited to bring Will in. It was a story I heard so many times in the years I worked for the RFDS. Patients would wait patiently until they knew a doctor would be available at the nearest weekly (or in this case monthly) GP clinic.

'I'll get some antibiotics into him when we get back.'

'Poor Will, hey Mum?' he repeated forlornly as she settled him into his car seat.

The landscape we drove through was reminiscent of the Middle East, and memories of a stint in Saudi came flooding back as desert scrub and bleached sand dunes stretched into the distance, not that I could see much; a long crack like a bolt of lightning ran across the windscreen of the Toyota Land Cruiser Michelle was driving. The splintered holes made it look like the car had been shot at. I'm exaggerating but only slightly.

'Tourists throw rocks at you,' she said.

Really? (No, of course not.)

'If I see a tourist with a trailer on a single-lane bitumen road I drop off to let them pass, they drop off too because they think they're being polite, only then they're driving with one wheel on the bitumen and they shower rocks at you.'

I made a mental note to remember that in future.

I'm not sure I believed Michelle when she told me that the area we were driving through had looked 'like Scotland' three years earlier.

'In 2010 we had ten times our annual rainfall. Those hills in the distance looked like the Scottish Highlands.'

There was no sign of any rain now.

'We're in drought now and we'll probably be in it for the next couple of years.'

Michelle's accent hadn't changed since I'd first spoken to her two years earlier. If there was any hint of an Aussie twang I certainly couldn't pick it. She still sounded like she lived on the outskirts of Crewe.

As we bounced along the dirt road towards Durham Downs, Michelle told me the unlikely story of how an English girl from the Midlands ended up living and working in the Australian Outback.

Like hordes of other young people from Europe, when twenty-something Michelle finished university she decided to postpone looking for work at home in favour of heading down under on a twelve-month working visa. By the end of May she'd travelled up and down the coastline from Sydney to Townsville, following the sun on her year-long odyssey. Bar work in Townsville dried up and she didn't fancy the prospect of spending winter in a cold, wet capital city so she contacted Visitoz, an agency that placed back-packers looking for temporary work.

'You'd go anywhere?' the interviewer asked, reading through Michelle's résumé, which contained details of a degree in sports science and experience on a farm and riding stables in the UK.

'I don't mind where I go,' Michelle said. 'The hotter the better as far as I'm concerned.'

'Do you know anything about cattle?'

Michelle hesitated. She'd grown up in a village on the outskirts of Crewe, an industrial town in the Midlands, sandwiched between Birmingham and Manchester and famous largely for its railway station. Part of the village onto a dairy farm, and as a child Michelle used to climb over the gate and lie down in a field full of cows, waiting to see how long it would take the curious creatures to circle her. (Ten minutes at most.)

'A bit,' she said.

The interviewer looked dubious, and with good reason. The

slight young woman sitting opposite him, with long blonde hair and a strong Midlands accent, looked more like a candidate for a bit-part on *Home and Away* than for the job on a remote cattle station he was trying to fill. What the interviewer didn't know was that Michelle was a talented gymnast, whose weight-training regime had given her a hidden strength.

'I can ride a horse,' Michelle added.

The interviewer shrugged. 'Well, there's a three-month job mustering cattle on Durham Downs, I suppose you might –'

'I'll take it.'

Michelle laughed as she remembered that interview.

'When I said I'd go anywhere, I had no idea where anywhere was.'

She arrived on the station at the start of winter, and loved it from the first moment.

'I kept thinking how much I would have loved being a kid out there, there was so much space to explore, especially if you had a horse.'

The head stockman at Durham was filling in before taking up a position as a manager on Glengyle (another Kidman property, further north) and Michelle was taken with his laid-back nature. Jon Cobb's slow confidence couldn't have been more different from the nervous energy that fuelled Michelle.

'I'd never met anyone like him. He poked along in a way that could sometimes frustrate me but he always got the job done. The more I got to know Cobby, the more I liked him.' There was no slow-motion dash across the paddock into each other's arms though. 'He didn't appeal to me at first. It was only once I got talking to him that I knew there was something there.'

Michelle had been planning to keep travelling at the end of her three-month contract but when Jon asked if she'd like to work with him on Glengyle, she changed her mind.

'I loved working with him so much I thought another three months wouldn't hurt.'

Three months turned into six and with her twelve-month visa due to expire Michelle was still working at Glengyle and still loving it. She knew she didn't want to go back to England so she went to Fiji, renewed her visa and returned as a tourist for another twelve months. Instead of doing the travelling she'd planned to do, Michelle stayed at Glengyle and worked for nothing.

'I couldn't get paid on a tourist visa so the company paid into my super instead.'

Those extra twelve months were enough for Michelle and Jon to realise that they had a future together. It began at Glengyle, 125 kilometres north of Birdsville, on 5500 square kilometres carrying 8500 cattle.

'It was a lovely place to live and work, great waterholes.'

The area around Glengyle was rich cattle-grazing country yet it was classified as arid: it received less than 168 millimetres of rainfall a year (six and a half inches). By way of comparison, in 2013 Sydney had 1344 millimetres or about fifty-two inches. The key was the flooding Georgina River that fed several lakes, including Lake Machattie, which covered an area of more than 900 square kilometres and provided one of the largest pelican breeding grounds in Australia.

Michelle combined raising children (three of their four boys were born while they were living at Glengyle) with working as a sports promotions officer in the local community. Then one day an opportunity came up that Jon had always dreamed of. The company asked him to take over Durham Downs – a prestigious property almost twice the size of Glengyle, carrying 25,000 head of cattle and frequently referred to as the 'jewel in the crown' of the Kidman cattle empire.

It was far more remote than Glengyle and there would be no opportunity for Michelle to continue her professional work in sport, but she was loath to stand in the way of Jon's ambition. Setting her reservations to one side, she began packing up their 'tidal wave' of accumulated possessions.

'The longer you're in a place the harder it is to leave.'

*

What struck me first about Durham Downs was the sheer scale of the place. The ground rose steeply from a causeway across an unseen river towards the staff quarters, spacious visitors' quarters and a storeroom. To the right was the all-important station kitchen, beyond the homestead was a schoolroom and on the crest of the rise were vast sheds with the words 'Durham Downs' picked out in letters several metres high on the roof. It was an impressive-looking operation.

We parked the car and walked across a yard of chalky ground, kicking up a fine powder of sharp sand and gritty dust. From what I could see, the dust seemed to coat everything it settled on, from boots, hats, bicycles and toys to dogs, horses and cars. The opaque, slowing-moving river was several metres below the homestead and hidden behind towering river gums and coolibah trees.

A patch of lawn in front of the house wouldn't have passed muster back in England but out here, in the vast Australian Outback, it was a startlingly green oasis that was clearly tended and treasured, and no wonder. It was hot and dusty now and this was the middle of winter. What would it be like in summer, when temperatures soared? For six months of the year, from October to March, the temperature at Durham Downs rarely dropped below thirty. In summer they could endure weeks of cloudless skies, with snakes an ever-present threat and flies a near-constant irritation. Days when the mercury was closer to fifty than forty weren't unusual, and that patch of green was the only colour in an otherwise parched landscape.

With Will safely tucked up in bed, Michelle showed me around.

The closest building to the station homestead was the school, originally a three-bedroom house from another station that had recently arrived on the back of a truck (expansion joints helped it arrive in one piece).

'This is such a fantastic schoolroom. You should have seen the old one. It had cardboard windows.'

The new building was partially renovated and Michelle had painted the room they'd decided to use for schoolwork bright purple and citrus green, changing her mind on orange when someone told her it encouraged anger.

'That's the last thing you want in a schoolroom!'

Inside it looked like most other schoolrooms, albeit a small one. There were affirmations and drawings, word maps and spelling bees, encouragement scores and maths problems pinned to the walls. The only things missing were children. Ten-year-old George was a lone figure in the corner, working on a computer. His younger brother, eight-year-old Joel, was tucked away in a separate room, also on a computer. Their oldest brother, Keegan, was away at boarding school.

'I know the boys don't know any different, but I think it's sad; it's not what I remember school to be. I was always so excited to see my friends and get involved in sport, play games. The only thing I didn't like was exams.'

My own childhood memories of a playground full of noisy children playing hopscotch, singing 'Ring a Ring o' Roses' or taking turns with a skipping rope were vastly different from the reality of school for Michelle's children.

'The biggest challenge is finding ways to motivate them. In a group, kids can motivate each other but when it's just the two of them it's so much harder.'

Even with material tailored to distance education students?

'It tells you to have a discussion. How can you have a discussion with one child? And brainstorming doesn't last long, we quickly send that to the floor.'

The tutor clearing away books was a cheerful young French Canadian who'd worked in China before applying for the job at Durham Downs. Andree-Anne had only been there a month and was clearly an intelligent girl with plenty of experience.

'George is a delightful child, kinaesthetic, and creative with his hands,' she said, 'and Joel is very quiet and always wants to please his teacher. Joel loves English and storytelling, whereas George prefers maths and science. And little Will is very considerate when he joins us in the afternoon.'

I was surprised to learn Andree-Anne wasn't a qualified teacher. The tutors Michelle employed didn't have to be; the only qualification they needed was a desire to live and work in a remote area. Try getting away with that in an inner-city primary school.

There'd been a hiatus before Andree-Anne arrived. One tutor left unexpectedly and Michelle had tried to convince herself she could manage, never mind that she was already employed by Kidman to do the bookwork (which included paying wages for a staff of twenty-plus) and was responsible for ordering all the station supplies as well as sorting out ad hoc human resources issues, plus she had to do the usual thousand and one jobs of any mum raising four young children (one of whom wasn't old enough to attend school at that stage).

'It was traumatic for the kids. The phones were going all day, messages were piling up and my mind was on other stuff. I just wasn't doing them justice.'

I reflected on how often as women we try to make ourselves 'manage' in a situation when most men would simply call in extra resources. In the end, Michelle advertised for another tutor and limited her own involvement to working with Will, who followed an e-kindy program on air every morning. She called into school every Friday to see how the children were doing.

I looked around the brightly coloured schoolroom, George sitting quietly on his computer. At first glance it seemed like an ideal set-up. George and Joel were extremely polite boys and they received one-on-one attention from their tutor, but the two-year age gap between them meant they were following different lessons. George was due to finish his unit in five weeks and Joel still had six weeks to go. What would they do then? And where was the school library? There was no music room, no gymnasium and no art room either. When did they ever run out and play footy in the playground with their mates or invite friends over for tea? There was so much we took for granted in urban areas.

Michelle had mentioned on the phone that a car rally came through Durham Downs one year, around the time of her oldest son Keegan's birthday. She'd organised a party to celebrate and his eyes shone at the crowd of people gathered to sing happy birthday while he blew out the candles on his cake. 'Mum, this is a *proper* birthday party,' he said. 'We've got guests!'

School camps were one way the children connected with others, participating in rare face-to-face lessons with other distance education kids. Michelle made sure they never missed a camp. When they lived at Glengyle she would drive them to camps at Mount Isa. The first year at Durham Downs she made the effort to take them back.

'It took us two days to get there, with an overnight stop halfway, and two days to get back.'

She didn't try that again. Instead she took them to Longreach, a mere 680 kilometres from Durham Downs, although the exceptional rain of 2010 added several hundred kilometres to their journey that year.

'The creek was flooded so I had to take a detour.'

Are you nodding as you read this? Thinking, ah yes, detours, they're so annoying aren't they? Think again. Michelle had to drive into South Australia, around Innamincka, then back around Noccundra to avoid the floodwater. After driving for eleven hours they reached the junction at the end of the bitumen on the road into Durham – a journey that would normally have taken an hour. She had four boys under the age of ten stuck in the car for an eleven-hour detour . . . *eleven hours*. And that was just the detour. Once they got to the end of the road they still had several hundred kilometres to go.

'The kids would have killed me if I hadn't got them there.'

Even without children of my own I had a feeling that journey would have been most parents' idea of hell.

And guess what? The journey back took even longer. Their car left the road on countless occasions, sliding off the camber as slippery wet mud stuck to the wheels and obliterated any traction. Pushing rocks under the wheels worked a few times but fifteen minutes from home they got well and truly bogged. Michelle radioed ahead and their tutor at the time, Kristy McGregor, helped keep the kids amused while they waited for help to arrive. The tractor that was sent to try to pull them free got caught in a sea of water, its wheels sliding helplessly through sticky mud as storms converged on Durham. In the end they were forced to abandon the car. Jon called Ballera and a chopper was sent to ferry them back to the station, but only after it had finished its rounds for the day.

'It was worth it to see the kids play footy with their mates,' said Michelle.

Is anyone still nodding?

'One of the hardest things about living in such a remote place is that the children don't have friends to play with. You never get along with your siblings like you do with your friends.'

Ironically, given all they space they had, Michelle was often hard-pressed to get the boys off their computers when they finished school.

'I bought them timers so they could limit themselves but if I'm busy in the office two hours can slip by and they'll still be on their computers. They're not silly, they take full advantage.' Keegan had bought his own iPad with winnings at gymkhanas.

'I couldn't really tell him not to use it.'

So it wasn't just kids stuck in a city with limited access to green space who weren't playing outdoors anymore – the seductive pull of electronic devices was pervasive. Michelle was no different from other parents worrying about the time her children spent on a computer. With Keegan away at boarding school when I visited, Michelle was hoping the school's strict rules limiting the use of personal electronic devices would break his dependency. (Any young person reading this will no doubt be horrified but my sympathy was with Michelle. Young people get outside, go climb a tree.)

When Jon and Michelle considered high-school options they had the choice of Toowoomba or Adelaide; Toowoomba was 1100 kilometres in one direction, Adelaide 1200 kilometres in the other. In the end they chose Toowoomba, partly because there were only seventy kilometres of dirt driving between Durham Downs and Toowoomba (as opposed to 700 on the road to Adelaide) and partly because Keegan was going to be boarding. Only a handful of boys from grade seven boarded at Sacred Heart or Rostrevor – the schools they were considering in Adelaide – whereas sixty or seventy boarded at that age in Toowoomba.

South Australia would have been easier in some ways because Jon had family there, plus there was a daily flight from the oil and gas field at Ballera, but there was also the weather to consider. Having lived in Queensland for most of their adult lives, Jon and Michelle couldn't conceive of retiring to somewhere like Adelaide, where winter temperatures regularly dropped to single digits.

'We thought if the children went to school in Queensland they'd be more likely to settle here. We don't know where we'll end up when we retire but the chances are it will be somewhere near the kids.'

If Michelle had lived in an urban area she would have had several high schools to choose from, some of them practically on her doorstep; living in a remote part of Australia, boarding school was the only option.

'I kept thinking we could have moved. We were only managing this place, we didn't have to stay here. It made me feel like we were bad parents, like we'd put business above family. We chose cattle over the kids.'

It didn't feel that way for Jon. He'd been sent to boarding school in Adelaide at the age ten, while his parents were working in Zambia, and he'd loved every minute. It was the norm for kids in remote areas.

Michelle didn't see Keegan for the first ten weeks. 'I had this overwhelming feeling of loss, like I'd lost a limb.' Then she drove to Toowoomba to celebrate her son's twelfth birthday and found he was having a great time, playing footy with his new-found mates and taking part in as many sporting activities as he could.

'I felt like slapping myself. He wasn't struggling in the way I was at all.'

The sun had dropped in the sky by the time we walked away from the schoolroom, intensifying the colour of the lawn that spread a carpet of green in front of the homestead. Crows and cockatoos called across the emptiness as we walked past a chook run with a lone rooster in it.

'A feral cat ate the hens.'

Michelle went in to fill the water up one day and found a massive cat in there. 'It got caught in the act. It was sleeping it off, paw hanging down over a dead body.' The cat had wormed its way into a chook pen strong enough to withstand attacks from dogs. 'Maybe it was our fault. Maybe it wasn't sturdy enough.'

I was tempted to tell Michelle the story of how Maggie, our domesticated dingo-kelpie cross, had eaten all our chooks but dingoes were a touchy subject on cattle stations so I kept quiet.

Beyond the chook run was a small veggie garden, the beds formed from rocks and bits of salvaged corrugated iron. Nestled inside were fat strawberries, flourishing silverbeet and rampant stands of bok choy. I spotted oregano, parsley, coriander and sage too, and basil growing in an old bathtub.

'I love my veggies. I was a vegan for many years, from thirteen until twenty. I wanted to change the world, didn't want animals killed.' I smiled inwardly. I'd been worried about being a vegetarian and Michelle was an ex-vegan turned cattle musterer? What strange and wonderful journeys life sends us on.

It was obvious that one small veggie garden could never have supplied a station with over twenty staff; most of their fresh veg came from Toowoomba, 1100 kilometres away. 'Fresh' was a relative term. Deliveries arrived once a fortnight, normally on a Friday (which more often than not meant Saturday or Sunday) and they never quite made it as far as Durham Downs. The closest drop-off point was the Santos plant at Ballera, fifty minutes away by dirt road.

'We must be like an annoying mosquito for the people who run that truck delivery business. Our order's not worth bothering with when they've got massive deliveries to oil rigs but we'd roll out the red carpet for them if they were to turn up here.'

Michelle would drive to Ballera to pick up the groceries on a Friday (or Saturday or Sunday) by which time they'd often been on the truck since the previous Wednesday – that's if they were there at all. On the last delivery an entire carton of Fuji apples was missing and they had to wait another fortnight to enjoy the simple pleasure of biting into an apple.

'Oh well, you can't win them all.'

I came across that calm acceptance time and again in the bush.

Mail was another thing they had to wait for; it arrived once a week on a plane that started in Port Augusta and took a week to complete the circuit of station properties. I hadn't spotted the airstrip so it must have been some distance away.

'The pilot flies over once to let us know he's there. He lands on the strip, we drive out, pick up the mail and drop off anything we want posted.'

If they missed hearing the plane fly over, the pilot might walk the mail down to them. If pushed for time, he would leave a small pile of post lying on the dirt.

Jon Cobb was finishing up for the day when we called into the office just off the main entrance to their homestead. A slight, lean man with an easy smile, his handshake was dry and firm.

'You used to work for the Flying Doctor?'

'That's right.'

'Reckon they should introduce a frequent flyer program for us.'

With over twenty staff, including young and inexperienced hands on short-term stays, accidents were inevitable, no matter how well regulated their work practices. Jon explained how, two or three times a year, someone might fall off a horse or roll their motorbike. If the accident happened at night they would light flares or shine car headlights along the strip to show the RFDS plane where to land.

'Cobby', as Michelle called him, had plenty of disaster stories to tell. He'd been airlifted a few times himself, once with a broken leg when his horse stumbled and fell on him.

'We've got a man who's fallen off his horse,' the lead nurse in the emergency department said as Cobby was wheeled in on a stretcher after a three-hour flight with the RFDS. The pain of his broken leg was nothing compared to the hurt to Cobby's pride.

'I didn't fall off,' he muttered. 'I was still on the horse when it fell over. There's a difference, you know.'

Joking aside, they both knew how fortunate they were to have access to medical assistance in such a remote spot. 'The RFDS paramedics at Ballera are brilliant,' said Jon. 'No question. We do everything we can to support that organisation.'

I knew from my research that farming accidents could prove fatal and they often involved children. According to a 2014 report released by the National Farm Injury Data Centre, thirteen per cent of farm deaths and thirteen per cent of farm injuries in the first six months of 2014 involved children aged fifteen and under. Dams were the biggest single cause of injury or death for children, followed by quad bikes then injuries associated with other farm vehicles, horses and machinery.

When it came to healthcare everyone on the station – adults and children – relied on the Santos clinic at Ballera, staffed by RFDS nurses on a regular basis and once a month by a doctor.

'The health clinic gets pretty busy on that day,' said Michelle. 'If it weren't for that clinic I don't think I could live here.'

Knowing what happened to her youngest son, I wasn't surprised.

*

It was mid-morning in late autumn 2011 and a clear day at Durham Downs. The RFDS plane had taken off a couple of hours earlier and the visit had gone well. There were enough staff at Durham to occasionally warrant a clinic held on site, and everyone appreciated a visit from the Flying Doctor. Ailments that had been saved up by those unwilling or unable to drive to Ballera to coincide with the monthly GP clinic had at last been treated and worries over heart murmurs, lumps, moles and medication had been addressed. It had been a good-humoured, lively visit.

The stock camp was out and a few staff had already left that day for a break, so smoko had been pretty quiet. The smell of coffee and chocolate muffins lingered in the warm kitchen as Michelle and some of the women working on the station helped clear up. Michelle glanced at the clock as she emptied the urn. She guessed the RFDS crew would be landing back in Broken Hill by now.

Suddenly there was a loud crash, followed by the heart-stopping sound of a child screaming.

Michelle looked at where her two-year-old son had been playing just moments before.

'Will!'

She spun round. Will was nowhere in sight. For a second she couldn't work out where the terrifying screams were coming from, then she looked across the room and saw the door to the meat house was standing open. In all the commotion someone must have forgotten to shut it. The room was full of equipment to cut and process meat. It was no place for a small child.

Michelle raced towards the meat room, heart thudding as the screams grew louder, and wrenched open the door. Will was lying face down on the concrete floor in a pool of blood. In a split second she guessed what must have happened. Will must have clambered up the fly screen – maybe to get a closer look at a frog or a lizard – and slipped off. As he fell he knocked over an old-fashioned sausage-making machine, and he and the machine hit the floor at the same time. The back of Will's head hit concrete and the lid of the machine clipped his forehead.

Michelle gently turned Will over and saw blood pouring from a gash on his forehead. The front of his skull looked like it had been pushed into his head and she could see white bone. 'Call Ballera!' she shouted as she picked up her son and carried him into the kitchen. She grabbed a clean towel to cover the wound.

Chris Belshaw ran the RFDS health clinic at Ballera. A senior nurse practitioner with years of experience in emergency work, Chris was half an hour away by helicopter and Michelle knew he could get there faster than the on-call plane from Broken Hill. Michelle cradled Will in her arms, doing what she could to comfort her son. 'Someone should radio Cobby,' she heard a voice say.

Jon was mustering cattle by motorbike somewhere out on the far reaches of the property when the call came through. 'Will's had an accident, get home quick,' he heard through static on his radio. He parked his motorbike on the flat and signalled the muster helicopter to land.

'Will's hurt,' he said, climbing into the helicopter and wondering what might have happened.

'I know,' the pilot replied. 'They're sending a chopper from Santos.'

They listened closely to radio traffic on the flight back, picking up that Chris Belshaw was on his way from Santos and that a plane was also being sent from Broken Hill with an RFDS emergency doctor. The pilot of the muster helicopter flew flat and fast. If both a helicopter and a plane had been despatched it had to be serious.

In the kitchen at Durham Downs Michelle felt increasingly helpless. Every time she looked at the gash on her son's forehead she felt responsible. A voice in her head kept repeating, 'You did this. It's your fault.'

She dressed the wound as best she could as Will clung to her, crying uncontrollably. Michelle paced up and down, trying to console him and keep her rising panic under control.

Time lost all meaning as she waited in the kitchen. It couldn't have been more than thirty minutes before Chris Belshaw arrived, but that was plenty long enough for Michelle to imagine all kinds of awful scenarios. When the solidly built man eventually stepped into the kitchen carrying an emergency response kit Michelle burst into tears. His open smile and compassionate face were deeply reassuring.

'What's happened here then?' Chris said calmly, in his lilting Irish accent.

Will kept clinging to Michelle as Chris checked the small boy's vital signs, and she looked away when his head wound was uncovered; a feeling of sick dread threatened to overwhelm her each time she spotted white bone.

Jon arrived just as Chris was finishing his examination. Seeing how distraught Michelle was he hid his concern at the sight of Will's injury.

'The plane from Broken Hill is on its way,' Chris said. 'They'll fly him straight down to Adelaide for surgery.'

Michelle felt a wave of panic.

'What about his head? What if there's serious damage to his skull? Or his brain?' Her arms trembled and her voice shook as she blurted out the question that had plagued her since she'd found Will on the floor.

'Let's just get him down to Adelaide, hey? They'll look after him there,' Jon said gently.

'Michelle, I need to take photographs for the surgeon,' said Chris. Michelle covered her eyes so she didn't have to look at the gaping wound on Will's forehead.

'It's okay, Will,' she said, suppressing a wave of sickness.

Chris forwarded the photographs of Will's injury to a plastic surgeon at the Women's and Children's Hospital in Adelaide, where they were gearing up to receive him, and the RFDS flew Michelle and Will down. Jon stayed at Durham to look after the other boys.

Will had refused to let go of his mum since she'd picked him up off the floor several hours earlier and he wasn't about to let go on the plane either. He clung to Michelle during the three-hour flight, forcing her to lie on the stretcher with him as the RFDS crew monitored his vital signs. When they arrived at the hospital Will was wheeled into X-ray then straight into surgery. It was only then that he finally let go.

It had been twelve hours since Will's accident that morning – twelve long hours that had stretched and elongated beyond measure – and now all Michelle could do was wait. She took a seat in the corridor outside surgery and sent up every prayer she could think of. Two hours later the surgeon appeared, with a prognosis that reduced her to tears.

'Your son should make a full recovery. And there's no sign of any further damage. It's purely cosmetic.'

Will bounced back pretty quickly. He went to the clinic at Ballera to have his stitches taken out and Michelle took the last few out herself with nail scissors a couple of weeks later. At the two-day Innamincka Races four months after his accident, where I caught up with Will and his mum, he was a bundle of joyful energy. Michelle took longer to bounce back.

'I don't know what happened to the sausage machine. I had it removed after the accident and I'm in no hurry to find it.'

Two years had passed since Will's accident and the meat house had been cleared of all its equipment. The small room Michelle showed me was quiet and still, opaque light filtering through dusty windows onto a bench where tender seedlings were being nurtured. The plant plugs were recovering from the shock of being sent by post on the mail plane; those that survived would be settled into raised beds in the vegetable patch outside.

Dinner that night was down by the creek, with a group of fishermen from Broken Hill. Visitors were few and far between in that part of the world and it was cause for celebration when they arrived – even 'poor Will' was keen to attend.

'We'll just pop down for a drink and say hello,' Michelle said.

Jon drove down with Joel and George and we followed with Will. The dusky light was fading as we bounced along a dirt track, heading for a spot on a wide bend in the river where the sand was soft and the yabbies plentiful.

The group camping down by the creek were regulars who came back year after year, fishing, kicking back and relaxing under the shade of centuries-old river gums. The 'Blokes from The Hill' were all members of the same footy team, they'd been mates for decades and when it came to camping they were renowned for their attention to detail. Some had been camping in the same spot every year for thirty-five years. They had been known to arrive at Durham Downs with candelabras, tablecloths and dinner jackets. This year, though, was special. The inclusion of wives had smartened up facilities even further, although the flowers on the table would have been there anyway, wives or no wives.

There was a campfire blazing on the riverbank, towels drying on twisted tree trunks, solar-powered fairy lights strung in the branches and a beer keg with its own tent. More fairy lights decorated a pop-up gazebo where long tables had been laid for dinner, and the canvas privacy screen that hung in the trees contained a shower – correction, *hot* shower. Water was pumped up from the creek by generator and heated by gas flame. The boys would normally have just pulled a tarp over a tree but the presence of 'the girls' had prompted a bit more cover. This wasn't like any kind of camping I'd ever known.

'Brought the basics again I see,' said Jon.

'Hey, good to see you guys.'

'How you goin'?'

'Who wants a beer?'

'Water's down a fair bit this year, Cobby.'

The campers had chosen a beautiful site under the trees. It was a great place to watch long-legged ibis pick their way along the shoreline and glimpse the occasional tortoise paddling through the water. Jon and Michelle knew the spot well. They would sometimes camp there with the children in school holidays, hitching their trailer tent to the back of the ute before driving it down the bumpy track to park it on the sand for a week. They may have been 1200 kilometres from the coast but there was fishing, swimming, water sports and boating to be had less than ten minutes away, with the added bonus that it wasn't such a long drive home at the end of the holiday.

Earlier in the day the fishermen had dropped a pair of stockings in the water (another benefit of having women along) with a tennis ball on top to keep them afloat and meat lures stuffed into the toes. The stockings had attracted bucket-loads of yabbies, some of which were now bubbling in a pot of boiling water.

I let slip that it was Michelle's birthday the following day and twenty or so people gathered around the fire launched into a chorus of 'Happy Birthday'.

'If we'd known we could have baked you a cake,' said Diane,

a retired teacher married to Ron, who at that moment was lifting the lid on the pot of yabbies.

They could have too. The mouth-watering meal they were preparing to serve had all been cooked in camp ovens – cast-iron pots with heavy lids – and somehow they'd turned out a full roast dinner. There was a choice of pork or chicken accompanied by cauliflower cheese, roast pumpkin, broccoli, roast potatoes, roast carrots, tomato pie and gravy. The night before they'd had rack of lamb and sticky date pudding.

'How are the yabbies?' Michelle asked.

Ron licked his fingers. 'Beautiful. We were on holiday in England two weeks ago and we ate at Rick Stein's place in Padstow.'

'Any good?'

'Yeah, but you know what? These yabbies beat the lobster he served.'

Another batch of cooked yabbies, which had been soaked in cooking water then hung aloft in a hessian bag so the cooling breeze could chill them down, was passed around.

'Here, try one.'

Someone handed me a fat tail and I peeled the brittle shell. I'd never tasted yabby before – I hadn't even known what they looked like – and even though I've never eaten at Rick Stein's award-winning restaurant in Padstow (or at Mollymook on the New South Wales south coast), I'd bet London to a brick those yabbies we ate were tastier than anything you'd get in *anyone's* restaurant. They were absolutely gorgeous.

There was general agreement though that it wasn't the best year.

'Jeez, some years you could drive in with a tractor and a bucket and just scoop 'em up!' shouted a voice from the darkness around the fire.

'Help yourselves, guys!' said Ron. 'Main course coming soon.'

Michelle took Will's car seat out of the car and settled it onto the sand by the river. Joel perched on a nearby log, George found an old camping chair and the boys sat chatting and laughing among themselves, adults ignored, car seat listing in the sand and Will's acute tonsillitis momentarily forgotten in the excitement of guests, an open fire, fairy lights, cans of soft drink and homemade Anzac biscuits.

'Hey, where's the big fella?' asked Peter, pulling the tail off another succulent yabby.

'At school in Toowoomba,' said Jon, who'd taken off his wide-brimmed Akubra to sit down for dinner (it was that sort of camp).

Peter turned to me. 'He's a real character. First year we came, Cobb had only just got here and Keegan can't have been more than five or six. He came riding in on his motorbike to check us out and he had a stick over his shoulder with a bag on the back like Tom Sawyer.'

Peter shook his head at the memory. 'He looked around the camp and said, "Hey, you've got a lot of stuff for one fella. Where can I park my bike?"'

Jon joined in the laughter. He was clearly proud of his son, of all of his sons.

'How's he doing at school?' asked Diane.

'He got a few tune-ups from the teacher in his first term but he's settled now. There were only four of them in his room and he had fights with two of them, said he was just sorting them out. Seems to be going all right now. He's good at making friends.' I may have been wrong but I could have sworn there was no hint of irony in Jon's voice.

I found a quiet moment to ask him later what he wanted for his children.

'I want them to get a good education. A lot of people who grow up in the bush are too insecure to go to the big smoke. We want our kids to be well balanced. If they want to be stockmen, we'll support them. If they want to be doctors, go for it,' he said, his voice confident and unhurried.

Jon was clearly happy in the work he'd chosen. 'It's normal for me to love what I'm doing. Some people come here just to earn a dollar and you wonder why. Since I was twelve, every school holidays I worked on a farm. I just thought it was interesting. Course, if you want to make a career in it you have to come up to these big places.'

The bush was a 'lawless place' in the early days of Jon's career. 'And I'm not that old,' he added. (I discovered later that he would be forty-eight on his next birthday.) The way Jon told it the environment was pretty wild back then, with none of the policies and procedures, licensing and liability insurance that governed operations today. 'The boss never knew where I was. I learnt more that way but I also made bigger mistakes.'

Macho competitiveness was rife, with men vying to prove who was top dog at local functions. 'I'd turn up to a pub in my twenties and people would get in a fight to prove who they were. Those days are gone.

'We've got more women working in the bush now,' he said, pointing to the female chopper pilot who mustered cattle at Durham. Under the right circumstances, Jon preferred to muster on horseback; like Michelle, he had a special affinity with horses.

'It's like learning the guitar. Anyone can play it but mastering it is different. Brilliant horses can teach you way more than you can teach them.'

Jon knew that Michelle's love of horses was part of the reason why she was prepared to live in such a remote spot. 'It's hard on her, living out here. She's stuck doing the cooking, cleaning and bookwork when she's a trained professional in sports science. I haven't made any sacrifices,' he admitted quietly. 'She has.' No doubt Michelle's love for Jon and the family life they had built together was another reason why she stayed.

'Come and get it!'

No one needed telling twice. Hungry campers gathered around the fire to fill their plates and the long camping table looked like a medieval banquet of food and drink.

'Who's got the apple sauce?'

'Mustard anyone?'

Michelle laughed. 'I keep expecting jugglers and entertainers to arrive!'

Once dinner was over, Will faded fast and he was soon curled up on his mum's lap, moaning softly to himself as the pain in his sore throat intensified.

Plates were scraped, food cleared, drinks finished and we said our goodnights. George and Joel politely thanked the campers for their dinner but Will was too poorly to do anything but groan.

Michelle left laden with gifts, jellybeans for the children and a huge jar of pickled onions for her – a birthday present that delighted her more than any expensive perfume or jewellery might have done.

'My favourite.' She grinned.

The year before the campers had left a jar of chilli-infused pickled onions and she hadn't been able to eat a single one. 'I can't handle hot food. It was torture watching Cobby tuck in to them every night.'

We left the campers grouped around the dying embers of their fire, with promises they would be back next year, and the remains of roast pumpkin on toast to look forward to for breakfast.

Michelle tucked a Tupperware of pork and roast chicken onto the back seat of the car.

'I've never had take-out from a camp before.'

*

At quarter to eight the next morning, ten-year-old George was standing in his pyjamas at the kitchen bench, breaking eggs into a mixing bowl.

'Happy birthday, Mum, I'm making you pancakes,' he said as Michelle appeared.

'We made you a card, only it's in the schoolroom,' said Joel, who was watching his older brother like a hawk. When Keegan went off to boarding school George took over the annual ritual of pancake making on Michelle's birthday. When George went, which could be as early as the following year, it would be Joel's responsibility.

'How many flours did you put in George?' Michelle asked.

'Four.'

'And four milk?'

'Yeah.'

'Eggs?'

'Six. You said times it by four but that looked too many so I only put in six.' George had already sprayed a frying pan with oil ready for the first pour and he was studiously beating the mixture with a whisk.

'That's the mark of a good chef, George, well done.'

'Shall I make some for Will, Mum?'

'Will's dying.'

Joel spun his head round and Michelle quickly corrected herself. The trauma of Will's accident had affected them all.

'Not really, Joely, he's just not feeling well,' she said.

Will was still in bed, struggling with the raging tonsillitis that had knocked the normally buoyant child off his feet. He'd woken up in the night crying with pain and Michelle had tried to get him to take another dose of antibiotics. Will had refused, choking with distress as he vomited bile and puss into the sink. He was so distraught Michelle wondered if she should call the Flying Doctor, something they only ever did as a last resort.

'I'll wait an hour,' she told Cobby.

Will settled eventually and Michelle slept fitfully on the floor next to his bed, hoping Joel wouldn't step on her if he got up in the middle of the night. She went through a checklist: the airstrip was clear – she'd faxed the weekly report through the other day – and the weather was good so if the RFDS had to land it wouldn't be a problem.

Thankfully, that didn't prove necessary and Will was sleeping at last.

The clock above George's head ticked towards eight o'clock and Michelle rang the schoolroom.

'The boys are making me pancakes for my birthday, would it be okay if they come to school at eight-thirty? It's taking a bit longer than we anticipated . . . Thanks, I'll send them in.'

'I asked if we could have the day off but she said no,' said George.

'Maybe we could finish at lunchtime?' Joel asked, hopefully.

'It's not a public holiday,' Michelle said firmly. I had to smile at their cheeky attempts to score a day off.

George poured the first pancake and tried to write his mum's name at the bottom of the pan. 'This is probably one of my best pancake mixtures,' he announced, proudly, gently lifting the pancake with a spatula before expertly flipping it over.

'Maybe you could go and get dressed while that pancake is cooking?'

Joel nipped back to his bedroom to get dressed but George

refused to leave his station. 'You have to watch pancakes all the time,' he said, intent on the task in hand.

'Do you want me to make you a coffee, Mum?' he asked. George was taking the responsibility for making his mum's birthday breakfast very seriously.

'Thank you, George, that would be lovely.'

He flicked on the kettle then carefully lifted another pancake out of the pan and added it to a growing pile in the oven, folding over a layer of tin foil to keep them warm.

'That's the best one so far, that one's for you, Mum.'

The phone went several times while George was making breakfast, once from a pilot in Toowoomba, where the station plane had been taken for an overhaul, and another time to check the strip was clear. Will called to his mum from the bedroom, there were calls to wish Michelle a happy birthday and I marvelled at her ability to stay calm in the face of so many demands on her time. No wonder trying to teach the children had proved impossible.

'Who was your favourite tutor?' I asked George.

'Miss Anne, she was good. And Diana.'

'No, Miss Rachel, she was the best,' said Joel.

'What about Miss Kristy?' Michelle asked.

'She was good but she made us work hard.'

Kristy was Michelle's favourite, a disciplined but fair teacher who stayed two years and who encouraged the children to strive for better results.

'She was a great home tutor.'

And after an unexpected stint of tutoring the children herself, salvation had arrived in the shape of the lovely Andree-Anne.

*

Michelle's brother was the only one of her three siblings from England who'd been out to visit. Her sister had been planning a trip to include a stay on the Gold Coast but took her children to Disney World instead.

'I told her she should have come here. It's pretty much the same as Warner Bros. We've got all the characters, and plenty of space and lots of animals.'

Quite apart from the horses they rode and the cattle they mustered, there was a donkey, a rooster, two pigs and the occasional pet kangaroo. Michelle reared an orphaned joey one time that grew into a huge male.

'He was my mate. He used to lie out on the veranda near the back door, stretched out big and proud.' She laughed, remembering the roo's attitude. 'He used to box you with his back legs, just for fun.'

One night the roo went up to the stables for a drink and Michelle found him the next day with a broken leg. She thought perhaps a horse had kicked him. 'It was so sad. I had to carry him down to Cobby to get him put down.'

Losing dogs was much harder. Growing up in England, Michelle always had a pet dog. 'A dog was for life, I never had to go through the trauma of losing dogs.' She struggled to accept that the dogs she tried to keep as pets at Durham Downs were often subject to fatal accidents.

'Cobby's working dogs have been run over, kicked by horses, bitten by snakes, you name it, and they've all survived.' Michelle's pet dogs hadn't been so lucky.

One young pup found its way into the storeroom and died hours later with a belly full of rat poison, despite Michelle's best efforts to save it. She poured salt water down the puppy's throat to make it throw up, even put it on oxygen, but it died all the same.

'You have to be your own vet out here.'

The water trick worked on their blue heeler, Nugget, when he ate rat bait and suffered no lasting ill effects, apart from bleeding from the nose for a couple of days, but 10–80 dingo bait killed him in the end.

'When Jon was baiting I told everyone, keep your dogs in your yards and don't let them come out for any reason.'

Dingo baits were lethal and for that reason were never placed anywhere near the homestead. Dogs could still die, though.

'If a dog licks a tyre that's driven over bait, or if it licks someone's boot that hasn't been cleaned properly, it will die.' She assumed that's what Nugget must have done. 'I hate baiting, absolutely hate it,' she said, vehemently.

I had to agree, although owning a dog that was three-quarters dingo no doubt coloured my opinion. I wasn't a farmer or grazier so it wasn't my livelihood at stake, but I often wondered how much damage dingoes did compared to eagles, foxes, feral cats and wild pigs. If you get rid of dingoes do you then have a feral-cat problem? Such views were probably hopelessly naïve so I kept my opinions to myself.

Snakes were another danger. The gardener's dog, Noo Noo, was fearless. He attacked the brown snakes and king browns that appeared at Durham Downs each summer and he was good at it too. One year he killed seven.

'You don't get between Noo Noo and a snake.'

Noo Noo's buddy, Monty, was a Rottweiler that also belonged to Lucy the gardener. It probably died because it was watching Noo Noo and wanted to join in. Monty died within hours of being bitten by a brown snake.

The saddest story Michelle told me was about Tyke.

'Tyke was a dachshund cross, the cutest puppy ever, with a deformed front leg that gave him a slight limp. He was my mate. He followed me everywhere.'

Jon had gone to a barbecue at another station and Michelle was staying at home with George, who wasn't well. When the puppy started yapping at the front door George went out to feed him scraps of meat. He ran back in screaming.

'Mum! A snake's got the dog!'

Michelle raced outside to find a python had Tyke in its mouth, its body wrapped tightly around the small puppy. She grappled with the snake and forced it to release the dog, but her desperate efforts to resuscitate Tyke failed. They'd had the little puppy less than two weeks.

'I was traumatised by that one. It was like after Willie's accident, I suffered flashbacks every time I closed my eyes.'

They buried Tyke at the bottom of the garden and Michelle vowed she would never get another dog.

'But there's something about them that I love so much.'

The wistful tone in her voice made me look up from my notebook. Michelle sheepishly admitted that she was looking at getting another dog. She was hoping to find an English bull terrier, a breed she fell in love with when she worked for a breeder in England.

'It might be a more robust dog.'

Living on a cattle station it will have to be.

*

Three generations of Sir Sidney Kidman's descendants gathered at Durham Downs in September 2011 to mark a hundred years of family ownership of five key stations, including Durham. The actual anniversary had taken place the year before, but the flooding rains that had bogged Michelle (the like of which hadn't been seen since the 1950s) forced the postponement of celebrations. The two-day event involved bronco branding, gymkhana rides and a massive celebratory dinner. Organising it must have been a huge undertaking.

'Jon and Michelle did an exceptional job co-ordinating the centenary, helping collect historical material from the stations and inviting past employees with a connection to the centenary properties,' said S Kidman & Co CEO Greg Campbell in RM Williams *Outback* magazine.

The following year, Michelle helped organise a Ladies Day at Durham, the brainchild of the children's tutor, Kristy McGregor. Around a hundred women enjoyed a jam-packed weekend of workshops, speakers, dinners, cocktails, health checks, therapy sessions and beauty advice. The oldest participant was ninety and women drove hundreds of kilometres to enjoy what for some was the first night they'd ever spent away from their children. The feedback was phenomenal and the event a huge success, in spite of flash storms on the first night that knocked out power, ravaged the marquee and trapped some of the participants on flooded roads.

'We had to go and pull out a few that got bogged.'

I tipped my hat to Jon and Michelle. Having worked as an event manager, I knew what weeks of unseen work were required to make such big celebrations run smoothly.

Jon was philosophical about what the future might hold.

'I don't want to be galloping around in the scrub looking for cattle when I'm sixty-five.'

That day was a long way off and he recognised that he was in the best place for now, pointing out that smaller family-owned operations were inevitably passed on to the next generation.

'It doesn't matter how good you are, sooner or later a son or daughter will always come home and take over the reins. If you want a career in cattle, you have to work on a big property.'

Michelle had no desire to move back to England; she preferred the idea of following the sun and retiring one day in Queensland. Besides, she now had four Aussie boys, five if you counted Cobby (as she sometimes did).

Jon's grandfather used to train horses in England, and his father once owned a stable block in 'the old country'. Jon had been back a few times with Michelle. He had fond memories of playing cricket on a summer's evening, when it was still light enough to see the ball at ten o'clock, but he wouldn't want to live there.

'If you've got a bit of dough England's a good place to be, but if you're a battler, working to make ends meet, you're way better off out here in Australia,' he concluded.

And if you're in agriculture, what better place to be than managing one of the jewels in the crown of the Kidman cattle empire?

Roma and Glenn Britnell

'Breland'
Woolsthorpe, 265 kilometres west of Melbourne, Victoria

I first heard about Roma Britnell through the Australian Rural Women's Award. I'd been looking for a family to interview in Victoria and Roma's name came up, first as winner of the Victorian award, then as national titleholder in 2009. A quick google search confirmed her high profile in the dairy industry and the fact that she and her husband Glenn had four children. Their youngest, Tessa, was ten years old at the time.

Their business looked extremely successful; they had not one but three dairies in the Warrnambool area milking over 1000 cows a day. I hesitated to call. As a writer I'm drawn towards people who overcome adversity – people whose success has come from battling the odds in some way and who struggle to achieve their dream. It's a bit like passing your driving test. If you sail through and pass the first time, there's no story; if you hit the kerb, stall the car and exceed the speed limit, there's a story, especially if you go on to pass the test on your fourth attempt.

I'm glad I picked up the phone. Far from being successful entrepreneurs covered in glory who had breezed through life on a dream run of ease and plenty, Roma and Glenn had to fight every step of the way to achieve their dream and they were still fighting to maintain it, with Roma leading the charge. In spite of her position as business manager of their Briland Farms operation, Roma cheerfully admitted she'd never wanted to be a farmer. 'I wanted to *marry* one,' she stressed.

Although both she and Glenn came from farming families, neither had inherited any part of their respective family farms. Roma's family lost their farm in a devastating drought that had tragic consequences and Glenn's father chose to sell the block he'd farmed since returning from World War II, without once thinking to stop and ask his son if he might like to be a farmer.

Time and again, Roma and Glenn were knocked back as the escalating price of farmland outstripped the pace at which they could save. 'You might as well settle for working on someone else's farm,' said bank manager after bank manager when they applied for finance. 'You will never be able to buy your own.' Never say 'never' to Roma and Glenn Britnell.

I met them in August 2013 at their farm three hours west of Melbourne, on a bitterly cold day of drenching rain and dense mud – the sort of day that makes you realise dairy farming is not for the faint-hearted. I was struck not only by their passion for farming but also by the extreme difference in their personalities. Glenn was a quietly spoken, introspective man whose affinity lay with animals and machinery; Roma was a voluble extrovert whose multi-tasking energy seemed unstoppable. Not only was she head of operations for their family business and a Nuffield scholar with multiple board appointments – including vice-president of United Dairyfarmers of Victoria – she was also a hands-on mum, closely involved in the local school, who did all the usual work of running a home and ferrying children back and forth to school and sporting activities.

Roma lived by the maxim she taught her children: 'Life is ten per cent what happens to you and ninety per cent what you do about it.'

When I heard what Roma and Glenn did about it, I knew they had a story to tell.

Family comes first

Reading through a wad of indecipherable interview notes and listening to hours of muffled tape recordings (*why* did I put the recorder in my pocket?) after I got back from visiting Roma and Glenn Britnell was like trying to make sense of a draft novel written in Russian and translated into Hungarian . . . with the pages out of order and no chapter headings.

I probably would have felt much the same way even if I had been able to make sense of my notes.

Roma Britnell was easily the most talkative driven woman I have ever met. Her energy was dazzling and unrelenting. Whatever Roma tackled – whether it was lobbying for a board position, driving her daughter to gymnastics, sorting out a problem with the tractor, turning turkey eggs, washing her mum's hair, stacking the dishwasher or negotiating the cost of repairs to a surge pulsator – she struck me as a woman on a mission to succeed.

Once I understood how challenging her journey had been, and how hard she and Glenn had had to fight for everything they achieved, I began to appreciate why.

Roma Britnell's father, Tom Hussey, came to farming relatively late in life, following a successful career as a stock agent with Dennys Lascelles. His wife Pauline pursued an equally successful career in nursing; she had already been appointed matron at the hospital in Nathalia, 230 kilometres north of Melbourne, when they met.

In his mid-thirties Tom was promoted to manager of the Dennys Lascelles office in Brighton, a beachside suburb of Melbourne, and he dipped his toe into the farming waters by investing in a hundred acres at Pakenham, in the foothills of the Dandenong Ranges.

The couple lived frugally as they saved money and raised a family of two boys and two girls, and the investment at Pakenham proved a sound one. Growing in confidence as a farmer, Tom sold the Pakenham property and bought 670 acres at Hawkesdale, three hours west of Melbourne. It was an area that offered more secure rainfall than the Riverina in New South Wales where Tom had been raised.

By the early 1970s Tom was able to resign from his job in Melbourne to pursue his dream, investing everything the couple had into the farm near the tiny town of Hawkesdale. The rich volcanic soil, high winter rainfall and flat, open pastureland made it ideally suited for running sheep.

Roma was four and a half at the time, their youngest child, and she had nothing but good memories of those days.

'Dad was at his happiest when he was farming. We'd get up early to go lamb marking before school and we all pitched in to help with whatever jobs needed to be done.'

We were sitting in the lounge room of Roma's brick bungalow, warmed by the ubiquitous wood burner, munching on cheese and biscuits and occasionally interrupted by the colt-like figure of Roma's ten-year-old daughter. Tessa would appear at random moments to perform a cartwheel or a handstand, her long red ponytail brushing the floor.

'Go and be upside down somewhere else please, Tess.'

'Would you like to meet my rabbit?'

In the blink of an eye Tessa was back, handing me a quivering bundle of fur.

'This is Flopsy, she has the fattest double chin.'

The conversation switched to pets, wild pigs and fox hunting, then to which of Roma's children was more in touch with nature (Austin the least, Vincent and Tom the most) and why Roma had let a male rabbit go that Tom brought home one night.

'You let it go? I knew it!'

'We had twenty-six rabbits, Tess.'

'I was just getting his confidence!'

'We couldn't afford to continue that particular breeding program.'

'Mum, do you remember when we saw that woman with a dog in her handbag?'

'No, I don't remember that.'

Flopsy hopped behind the sofa and I tried to steer the conversation back to Roma's childhood. It wasn't easy and I got the sense Roma wasn't one for sitting down much. Our interview seemed to have the unfortunate consequence (from my point of view, anyway) of turning her mind into a trapped animal racing back and forth in a bid for freedom.

Maybe she didn't like being interviewed? There was no hope of keeping track of the welter of information, so I put my notepad down and slipped the all-too-obvious tape recorder into my pocket (and, oh boy, did I regret doing that) hoping to put Roma more at ease. Then, as the conversation continued, I realised it had been nothing to do with nerves; it was more to do with the story she was about to tell me.

'Dad always used to whistle children's songs when he fed out so I could sing along.'

At seven, Roma would steer her father's ute through their flock of sheep on the flat plains at Hawkesdale while her long-limbed father pulled hay off the back, whistling 'Bingo Was His Name-o!' and 'Where is Thumbkin?'. The farm prospered, her father flourished in his new life and the picture looked rosy.

Then the 1976 drought hit. It wasn't the worst drought on record but it was a severe one and it came just three years after the extreme ten-month drought of 1972–73. It also followed a less than average yield the year before.

The drought hit hard in coastal areas west of Melbourne – including Hawkesdale – and the prolonged dry wiped out stocks of feed. Borrowing money to buy more feed when the annual rate of inflation was over thirteen per cent wasn't feasible for small farmers like the Husseys so they clung on, searching the skies for signs of rain.

The tipping point came when the market for sheep suffered a severe drop in price. 'Victorian Farmers Flayed by Drought and Falling Prices' ran the headline in the *Canberra Times* on 12 May 1976. The cost of trucking sheep to market overtook the price those sheep would fetch and farmers faced an agonising choice; with no rain in sight, those who couldn't afford to buy feed had to either watch their animals die a slow death from starvation or shoot them en masse. Tom Hussey chose to shoot his sheep.

'It must have been heartbreaking for Dad.'

'How old were you?'

'I was eight. My brothers and sisters were eleven, twelve and thirteen.'

With the flock gone, the farm locked in drought and no money coming in, 41-year-old Tom had been forced to abandon farming. Reluctantly, he took a job as licensee manager of the pub in Warrnambool, forty kilometres away, and Roma's mother was left to close up the house and the farm. The children were enrolled in new schools, the farm was leased at a peppercorn rent and six weeks later they joined their father in Warrnambool.

'We packed everything up, moved into the pub and started a new life.'

I was burning with questions. How did her father cope? Did Roma know the sheep cull was happening? How did they feel about the move? How long did it take to settle? What about her mother? How did she cope with such terrible events?

'I don't remember,' Roma said when I began to ask.

Surely that couldn't be right? My sceptical disbelief was short-lived when Roma revealed why she recalled so little of the trauma that must have surrounded the forced cull. What happened next overshadowed everything else. Six weeks after they arrived in Warrnambool, in July 1976, Tom Hussey suffered a cerebral aneurysm and died, leaving a widow and four children. Roma was eight years old.

'It was a horrific time. Even now, forty years later, I can still get upset thinking about it. One of my first thoughts after Dad died was that, when I had children, I would call my first son Tom. It wasn't a fashionable name in those days and I remember thinking what a shame that was.'

Roma's mother put her grief to one side as she struggled to provide for her children. Knowing how happy they'd been on the farm she used the insurance money to get out of the hotel, go back to Hawkesdale and restock.

'Everyone saw Mum as this strong, energetic woman. What they didn't see was the grieving widow who fought during the day and collapsed at night.'

No wonder she collapsed. Astonishingly, Pauline's efforts to re-establish the farm were thwarted by unscrupulous competitors who cut fences and boxed up sheep, mixing stock to deliberately confuse the inexperienced widow. As a single mother coping with grief, she was an easy target for those trying to force a sale. Within twelve months of Tom's death, she was forced to admit defeat. Overwhelmed by the task ahead of her, Pauline sold the family farm.

And suddenly Roma was crying. As she apologised (which wasn't necessary but I felt for her, I was a total stranger and we'd only just met) I caught a glimpse of the impact that such tragic events in her early childhood must have had. Roma's energy and 'unstoppable drive' had its origins in heartbreaking disaster.

The loss hit everyone hard. Looking back, Roma thought the typical teenage misbehaviour of her older brothers and sisters might have masked a lot of anger. 'I think my older brothers were trying to be men, trying to take Dad's place. They were burying their grief.'

The world Roma knew collapsed around her. 'Mum would be so strong during the day then she would fall apart at night.' Night after night, Roma would climb into bed with her mother and try to console her. 'The worst part was being forced to sell the farm. Mum wasn't angry and she wasn't defeated, she was just so very sad.'

Small wonder that Roma was such a determined and driven farmer herself.

She broke the tension with a short laugh.

'I never set out to become a farmer. I wanted to *marry* a farmer, I didn't want to *be* one.'

Two years after her father died, when she was ten, Roma decided she wanted three things out of life – she wanted to be a nurse like her mother, she wanted four children and she wanted to marry a farmer.

She achieved all of her ambitions, although her husband Glenn wasn't a farmer when she married him; he was a sheep shearer.

'And the one thing Dad told me was, whoever you marry, make sure he's not a shearer!'

*

Tessa's bedroom was typical of most ten-year-old girls, decked out in pink and purple like a fairy grotto and covered in posters of horses, dogs and her current favourite boy band, One Direction. There were clothes strewn on the bed and a bookcase with manuals on horsemanship and well-thumbed copies of Enid Blyton. The only difference with most other children's bedrooms lay at the bottom of her wardrobe, where twelve turkey eggs nestled in an incubator. The eggs had to be turned every twenty-four hours until they hatched.

'If they hatch in the night will there be turkeys running around the room when I wake up in the morning, Mum?'

Roma shook her head. 'I think we'll have to move it before then.' Roma knew that if the eggs weren't fertile they would explode and the smell would be unbearable.

'Have you fed the lamb?'

Tessa's eyes widened and she sprang to her feet.

'You'd better hurry up before it gets dark.'

Tessa performed a handstand in the corridor on her way to the kitchen, a feat she enjoyed so much she repeated it in the living room and all thought of feeding the lamb was momentarily forgotten.

'Tessa Monique, what are you doing upside down again? Go and do that somewhere else. And don't forget that lamb.'

Tessa righted herself, filled an old beer bottle with fresh milk, added a rubber teat and slid open the glass door separating the kitchen diner from a covered extension. Feeding a lamb sounded like fun so I joined her.

'Help yourself to a pair of boots, you'll need them,' Roma shouted. Wet-weather gear hung in soggy clumps under the plastic roof and muddy boots lined the walls. 'And Tessa shut that door before all the cold air comes in!'

The glass door was covered in a pink scribble of chores – *clean out calf shed, fence posts, water pipe number seven, fix calf feeder* – faint blotches of marker pen still visible where completed jobs had been rubbed off. Tessa slid the glass door closed and wriggled her feet into a pair of flats.

'Tessa Monique, those are my shoes!'

Tessa skipped away, pretending she couldn't hear, running across the lawn in a series of frisky leaps. She looked like a young colt. I followed (like a lumpy koala), ducking under an arch formed by pine trees and past a woodpile where a huge turkey was scratching in the dirt. It answered Tessa's accurate call with a throaty gobble of his own.

'Mr Gobbledoc nearly died you know.'

He was clearly luckier than most turkeys. When the turkey picked up an infection the concerned Britnells took him to see a vet – the first time the surprised man had ever been asked to treat a turkey – and Mr Gobbledoc displayed so much personality when he recovered that he was saved from the knife.

'We decided to keep him as a pet.'

The turkey followed us across an open paddock, gobbling at Tessa.

'And this is Ranga, he's nice but they're a bit mean, I think because they have piglets.'

Tessa pointed towards three large pigs, two of which had litters, and Roma joined us to reveal one litter was destined for a twenty-first birthday party and the other for a community event being held to raise money for a new roof on the local hall. (Mr Gobbledoc was one very lucky turkey.)

There were caravans for staff, towering rolls of silage wrapped in plastic and a substantial chook run with space at one end for the orphaned thirteen-week-old lamb Tessa had come to feed.

'Hello Jimmy.'

She pushed a bottle through the fence and Jimmy sucked

greedily, emptying the contents in seconds. The lamb had to be fed three times a day: before school, after school and again before bed. Tessa would have to come out once more on that cold, wet night for his final feed.

'All of the children have done it,' said Roma. 'It was character-building for them to get up early in winter and go out late at night to look after animals that were totally reliant on them.'

Tom, Austin and Vincent no longer lived at home but they'd all taken it in turns to look after orphaned lambs from the adjacent sheep farm. Vincent had loved it so much he was left in charge of twenty-seven lambs one year. Roma suspected his eagerness had as much to do with promised dollar signs as with animal welfare.

Roma showed me through the dairy, an impressive semi-automated operation with a closed system that guarded against bacterial infection and dropped the temperature of the milk down to four degrees within four hours of milking (I hope I got that bit right). She opened a tap and fresh milk gushed into a jug.

While we'd been looking at the dairy Tessa had collected eggs from the chooks. She'd checked that budgies Abigail and Henry had water, she'd put Flopsy away before the rabbit could eat any more of the spare mattress behind the sofa in the living room, and she'd fed the cats Albert, Scratch and Snowball.

'Oh, Matilda!' she said suddenly. 'I need to walk Matilda, I forgot!' She dropped the empty bottle and ran towards a horse standing by a fence at the bottom of the garden.

'Don't worry, Tess, she won't need a walk tonight. Come on, it's getting dark.'

*

Glenn Britnell was such a quietly reserved man that I barely noticed him slip into the lounge room while Tessa was cartwheeling and Roma was showing me family photographs, rattling through a barrage of details like a firework display.

'It's bedlam in here,' a voice said.

I looked up at a laughing man with dark curly hair and a beer in his hand.

'Meet Glenn,' Roma said, smiling back at him before launching into details of a quote for a new surge pulsator and quizzing Glenn closely on why it was needed.

'If it doesn't fix the problem why are we spending 2000 on it? And I had a call from the tractor man, no, not the old one, the new tractor, he said about 89,000, 3.9 per cent interest with 30 per cent upfront and a balloon over . . .'

'Hang on, slow down.' (I wish I'd thought to suggest that.)

Roma ran through the figures again – slowly – and Glenn nodded then turned to his daughter.

'Have you done the budgies? And the lamb?'

'Yes, Dad.'

'Good girl.'

It was fascinating to watch how Roma and Glenn interacted.

'We are ridiculously polar opposites,' Roma admitted, laughing. 'I'm right out on the spectrum of extrovert and Glenn is right out on the spectrum of introvert.'

How did two such wildly different people meet?

'You tell the story,' said Glenn.

'No, you,' Roma countered.

Glenn looked hesitant. 'Tess, you tell the story,' Roma said.

'Well, you were camping, and your boyfriend and Dad were best friends and they were in a car crash and your boyfriend died and then Mum was boyfriends with Dad and that's all I can remember.'

She looked shyly at her parents to see if she'd got it right.

'Sort of true,' Roma answered, smiling.

*

Glenn's father was a soldier settler who was awarded a block of land near Macarthur in southwest Victoria after World War II. The quietly spoken man was one of many soldiers who fought in the Far East. He was imprisoned for a time at Changi.

'The only time we ever heard stories was when they had a reunion. It's not something he ever spoke about.' The other thing Glenn's reserved father didn't speak about was farming.

Like most soldier settlers, Jim Britnell started with dairy. The farm was on a good block of land and over the years he expanded, incorporating sheep and a few beef cattle. Glenn was the couple's last child, born when his father was already fifty and his mother in her late forties. A quiet child who took after his father, Glenn was often left in charge of the farm when his parents went away. From his mid-teens onwards (by which time his older brothers and sisters had all left home) Glenn also earned money as a shearer, going away for weeks at a time on shearing camps. He loved the solitude and concentration of the work and would happily shear for hours at a time, as focused as an athlete.

He came back from one trip to hear his parents announce they had decided to sell up and retire. Glenn was sixteen at the time and it never occurred to him to talk to his father about his own ambitions to farm, nor did he feel any resentment about his father's decision to sell the farm without consulting him. He simply took to the road and embraced life as an itinerant shearer.

The car crash that killed Glenn's best friend, Mark, happened three years later, in July 1983, the night before nineteen-year-old Glenn had been due to leave on a twelve-month odyssey around Australia. Glenn was devastated by the accident but, knowing his mate would have wanted him to continue, he set off a month later with another mutual friend, Dave. Several weeks later they ended up in Alice Springs, where Roma was meant to be studying.

Pauline had sent her sixteen-year-old daughter to stay with her older sister Cathy – a sensible girl engaged to a local policeman in Alice Springs – in the hope that it might help her get over Mark's death. Pauline was worried that Roma had been so upset by what happened that it was affecting her studies. With exams coming up, she thought a period away from all the talk might help.

Roma's sister worked as a nurse, and the nurses' home where she lived was quiet during the day. It was the third-term school holidays, September 1983, and Cathy wasn't due back for several hours. With her books open on the table in front of her, there was nothing to distract Roma from the task in hand; even so, she was finding it hard to concentrate.

The knock at the door was a welcome distraction and Roma opened it to find two lads she knew standing on the doorstep. Dave was an old school friend of Cathy's and the diffident shearer with straw-coloured hair and whippet-thin legs who had gone to school with Roma's brother was also a familiar face: Glenn had been Mark's best mate.

'G'day guys,' she said with surprise. 'What are you doing here?'

'Up for a couple of weeks, heading out to the rock if you want to come?'

Roma jumped at the chance to ditch her schoolwork and invited the boys in while she grabbed a few things.

'I'll just leave a note for my sister.'

The adventurous sixteen-year-old tore a page out of one of her exercise books and scribbled on it, 'Gone to Ayers Rock with Dave and Glenn. Back soon.'

Back soon? I once flew to Alice as a naïve tourist, thinking I'd be able to hire a car, drive out to Uluru (or 'the rock' as everyone called it then) and be back in time for tea. No chance. I had no idea how far it was from Alice Springs and it sounded like Roma didn't either, or if she did she didn't care.

This was before the days of mobile phones so there was no way for Roma to stay in touch with her sister. The trio eventually got back three days later.

'Hi Cath, I'm home.'

Roma's older sister grabbed her by the arm and spun her round.

'I am going to *kill* you!' she thundered. 'Where the hell do you think you've been?'

'I left a note.'

'A note that said you'd be back soon and that was three days ago! You're lucky I didn't call the police! I was meant to be looking after you. Do you have any idea how worried I was?'

Cathy then directed her anger towards Glenn and Dave. 'And you idiots should have known better than to take a sixteen-year-old girl away for three days!' she shouted at the hapless pair.

'Nothing happened, you know,' said Roma.

'I should bloody well hope not!'

Cathy's mixture of relief and anger was fuelled by several late-night phone calls she'd fielded from their mother, having to invent all sorts of excuses why Roma couldn't come to the phone at that precise moment.

Roma grinned at Glenn behind her sister's back. She hadn't been lying when she said nothing had happened, but she'd had a lot of fun on the three-day camping trip. Spending time together had helped them both cope with the tragic accident and the shy nineteen-year-old shearer, who would only talk to Roma under cover of darkness, had captured her attention.

Roma went home and pinned a photograph of Glenn onto her blackboard at school.

'Here.'

She passed me a movie-star image of a lean young man smiling at the camera. He wore boots, a blue singlet and shorts, his arms and legs deeply tanned, eyes screwed up against the sun and a blade of grass hanging from the side of his mouth.

Two weeks later, Glenn cut short his trip.

'We've been joined at the hip ever since,' said Roma.

They married at the Catholic church in Warrnambool, in March 1987, and started married life together in Hawkesdale, the scene of some of Roma's happiest childhood memories.

'I think I was trying to recapture what we'd lost,' she said, referring back to the forced sale of her parents' property.

The first few years followed a familiar pattern. Glenn would go away shearing for weeks at a time and Roma worked every shift she could at the local hospital, where she was in her final year of training to become a nurse.

'I worked every weekend and never took holidays because we were determined to save enough money to buy a farm one day.'

Roma furthered her nursing skills with postgraduate cardiac training, knowing it would always give her something to fall back on, and they started a family pretty much straightaway. The imminent birth of their first child prompted some stern advice from Pauline, Roma's mum.

'Right,' she said to Glenn. 'You need to buy a house now.'

While Roma was in hospital waiting to give birth, Glenn obliged and six weeks later Roma and the new baby (Glenn Thomas, known to everyone as Tom) moved into a house with eight acres of land. Typically, Glenn was away shearing at the time.

Five months after Tom's birth, interest rates started to climb and Roma went back to work to help cover the cost.

'Within twelve months our interest rate had shot up to eighteen per cent.'

While interest rates stayed high, Roma and Glenn embarked on what Roma called 'an adventure'.

They rented their house out and moved to Deniliquin, a small country town just across the border in New South Wales, where they reduced their outgoings by living in a caravan on the banks of the Edward River, a branch of the mighty Murray.

'People asked us why we were doing it. It was partly to save money but it was also just because of the adventure. We had a great time, we loved it.'

The rural peace and quiet of Deniliquin suited the whole family. 'We swam in the river every day and had barbecues around a campfire with friends every night. It was perfect.'

Glenn took all the work he could get shearing and Roma took a job in the accident and emergency department of Deniliquin Hospital, although she cheerfully admitted she was hopeless when it came to any crisis that involved her own child.

'Tom suffered an asthma attack one night, it was so bad he started vomiting and I panicked, which is the worst thing you can do.'

Glenn was between shearing jobs so he took charge that night, pushing Roma out of the way as they carried Tom into casualty.

'I was a mess behind him.'

The doctor in emergency thought it hugely funny that the normally controlled and capable nurse he worked with had been so completely undone. He took one look at Roma and triaged the situation. 'Right, Valium for her, Ventolin for the boy.'

Mindful of the need to save for the farm they one day wanted to buy, Glenn and Roma invested in another property in Deniliquin and again they rented it out.

'We were quite business-focused even then. Our goal was to buy a farm and we worked our butts off earning enough money to get there.'

Their second son, Austin, was born four years after Tom, by which time they had moved back to Hawkesdale, setting up the closest they could get to a farm with a menagerie of animals.

'We had turkeys, chooks, a cow, a few sheep, plus cats, dogs and rabbits.'

Tom had a natural affinity with animals, just like his father, who always knew where the plovers that mated for life nested and made sure he ploughed around them.

'Tom had an idyllic life there.'

And what about you, I wondered? I could understand why Roma had tried to recapture what she'd lost as a child, and it sounded like she had created a rural idyll for her young family, where Tom was clearly in his element. But it can't have been much fun for Roma. With Glenn away shearing for up to seven months of the year – in blocks of two, three or sometimes six weeks – she was left to milk the cow, feed the chooks, look after the sheep, raise two children and somehow find time to work three days a week as well. Was it tough?

'We both knew it had to be done if we wanted to save enough money to realise our dream.'

(I took that as a yes.)

The dream proved elusive. No matter how hard they worked, the prospect of owning their own farm was always just out of reach.

'We saved $10,000 and went to the bank to ask if we could borrow enough to buy a hundred acres. The manager just laughed.'

The couple put their heads down and saved $50,000. They

bought forty head of cattle and put them on a lease block to prove they knew what they were doing and approached the bank again. Again they were knocked back.

'They gave us a flat no.'

Refusing to be beaten, they worked harder, spent less and managed to accumulate $80,000, and still the bank refused to loan them any money.

'It was so disheartening. We worked our butts off and it was never enough.'

The dispirited couple knuckled down and kept saving.

Glenn cracked the top on another beer and Roma ducked into the kitchen to check on dinner.

'Don't forget Jimmy!' I heard her shout to Tessa.

'You lived in the Hill?' Glenn asked as he handed me a bottle. I nodded.

'I sheared up there in the wet one year, just before Vinnie was born.'

So with a third child on the way Glenn had still been shearing and Roma must have been working part-time as a nurse. The longer that went on, and the bigger their family grew, the more elusive their dream must have seemed.

I wondered how close they ever came to giving up? Roma's mother had been forced to relinquish her dream and, from what Glenn told me, he and Roma faced a similar scenario.

*

The sky above Ivanhoe was dark with cloud. Glenn put his clippers down and wiped the flies from his face. He'd never seen so many insects. The air was humming with them. He could normally concentrate when he was shearing, like an athlete in a long-distance race, but the mosquitoes eating his face were impossible to ignore. He swatted them away.

'Hey Glenn, I reckon those mozzies could pick you up and carry you off'.

Glenn laughed and grabbed another sheep. No matter how much he ate he could never keep weight on when he was shearing. His thin body was down to ten and a half stone. He wiped the back of his arm across his eyes, cleared away the sweat and clouds of mosquitoes that obscured his vision, and flipped the sheep onto its back.

It had been a wet year. In January he'd been shearing in forty-degree heat at the back of Menindee Lakes, between Broken Hill and Wentworth. The yards had been full of water and shearers were barely able to lift the flyblown sheep. Then a downpour of ten inches in less than two days stopped all shearing; for a while the shearers were trapped on the property. Glenn got out by walking sixteen kilometres to the main road, most of it through waist-deep water.

Autumn was proving just as wet as summer. The team he was working with in Ivanhoe – 800 kilometres away from Hawkesdale – had started shearing at six o'clock that morning to escape the killing humidity.

He was looking forward to getting home.

Austin, his three-year-old, had hidden behind a door and refused to come out when he'd tried to say goodbye before his last trip north. The frequent absences unsettled the small child. What's more, Roma was heavily pregnant with their third and Glenn knew the burden of looking after the house, the animals and the children fell squarely on her whenever he went away. He was glad his next job would be closer to home. When he finished in Ivanhoe he'd be working at his brother-in-law's sheep farm, just two hours from where he and Roma lived.

'We're coming with you on the next job,' Roma announced when Glenn got back from the stint in Ivanhoe. She'd been forced to go on maternity leave several weeks before and she was starting to crawl the walls.

'I thought you were due any day?'

'Couple of weeks yet,' she said, mentally crossing her fingers.

'What about Tom?'

'He can stay with Mum. I haven't seen Mark and Clare for a while and you know Austin hates it when you go away.'

Roma's mind was made up and Glenn reluctantly had to agree.

As soon as shearing got underway, Roma pitched in to help. A day's physical labour as roustabout in the shearing sheds, picking up fleeces and throwing them onto the skirting table, had the inevitable consequences. At four o'clock the next morning Roma shook Glenn awake.

'I'm in labour.'

'Are you sure? Try going back to sleep,' he mumbled.

Roma timed the contractions at five minutes apart and woke him again.

'We have to go home.'

It was a three-hour drive back to Warrnambool across the Grampian Mountains, where there would be mobs of unpredictable kangaroos and no mobile phone coverage.

'Let's go to the local hospital.'

'No! I want to have all my babies at Warrnambool, you know that.'

Roma's brother Mark was no more successful at persuading her to stay, nor was his wife, Clare, in spite of the fact that she was a nurse and had been a friend of Roma's since the age of five. Roma refused to listen to reason.

Glenn insisted on a cooked breakfast, hoping the delay would make Roma come to her senses and realise the futility of driving three hours across the Grampian Mountains, but even that failed to deter her.

In desperation he blurted out, 'We haven't got enough petrol.'

'Glenn, will you please get the car? If we set off now we'll make it.'

Glenn bundled Austin into his car seat, urged their three dogs to jump into the back and they set off, driving through a dark night with no moon to guide them.

'How many minutes now?' Glenn asked, clutching the steering wheel as a shadow on the side of the road loped away.

'Three,' Roma replied.

That was enough for the normally placid Glenn.

'You know what? I'm not doing this.'

He turned the car around in the middle of the road and drove back to the tiny hospital in Stawell, where a surprised midwife answered his urgent knocking. Roma just had time to call her sister-in-law and ask her to bring a video camera before Vincent was born. Four hours later they were on their way back to Warrnambool and Roma woke up in hospital the next day feeling like she'd been hit by a truck.

With another shearing job booked in Queensland, Glenn couldn't even stay until Roma left hospital. The excitement of Vincent's birth was set aside and he picked up his shears and swag and headed north.

Three weeks later, Glenn stood in the doorway of their house in Hawkesdale, watching Roma nurse Vincent. He'd come back from the exhausting job in Queensland to find Roma had been in hospital with a chest infection. Seven-year-old Tom had done his best to tend the animals and help look after his younger brother while his mum recovered and he had dark circles under his eyes. Three-year-old Austin was in tears and no one looked happy. It was the final straw for Glenn.

'We're working ourselves to the bone and for what? We are never going to achieve this crazy dream of buying a farm. We're trying to raise a family and I'm never home. Our children won't love us for it. Look how Austin reacts every time I leave! We have to forget it. Now.'

Roma's stomach lurched.

'Glenn, you're a shearer and I'm a nurse. Are you suggesting you give up shearing? We can't afford for you to stop work and I can't go back to nursing for at least another three weeks. What's your plan?'

Glenn didn't have one. All he knew was his family was suffering and he'd had enough.

Two days later he picked up the morning paper and saw a job advertised for a trainee dairy farmer.

'What do you think?' he said, pointing to the ad.

Roma read through the brief description. 'It says you'd have to go back to school and study.'

The prospect was a daunting one. Glenn had left school at sixteen and never looked back; he wasn't a natural student.

'It would mean a big pay cut as well,' he added.

'We said we didn't want to be dirty old dairy farmers.'

'Never taking any holidays.'

'Up to our knees in shit every day.'

Glenn nodded. He drew the paper towards him and read the ad again. 'If we forget about sheep or cattle and concentrate on dairy,' he said, 'we might still be able to afford our own farm one day.'

He picked up a pen and circled the ad.

'I'm going to apply.'

*

It was the turning point they'd been waiting for. Glenn embarked on the twelve-month course with enthusiasm, encouraged to find the tutor was someone who used to live next door to his family in Macarthur where he grew up. At weekends he worked for a nearby dairy farmer, earning the princely sum of eight dollars an hour. In 1996 that was about half the average wage and it had to feed and maintain a family of five.

Putting the kids in childcare would have cost more than Roma could have earned part-time as a nurse so she stayed at home to look after them. While Glenn studied, Roma read as much as she could about an industry that was new to them both, and what she discovered surprised her.

'It was like the best-kept secret. Here was a rural industry in Victoria that was scientifically sound, stimulating, environment-ally proactive and it offered us a career pathway into agriculture. We looked at it and thought, okay, we may not be able to own our own farm but we can still get the benefits of the lifestyle.'

The farmer Glenn worked for was newly single. Having just gone through a divorce, he was happy for them to stay at his house whenever he was away. That gave them the chance to experience the reality of family life on a dairy farm.

'We loved it.'

Every morning Roma and the children would go out on motorbikes to get the cows and bring them in for milking, all the time watching and learning. Administering medication or intravenous injections was something Roma could easily relate to and the similarities between nursing and dairy farming fascinated her.

'You have to closely monitor a cow's post-partum health and nutrition, and it's fascinating how similar the treatments and disease processes are for infections like mastitis. Cows are like women when they're having babies.'

Members of the consortium that owned the farm were quick to spot Glenn and Roma's initiative and soon asked them to look after Riverside, another farm the consortium owned. In the first year the couple milked 180 cows, in the second 240 and it was only in the third year, when they expanded to 350, that they had any help.

By now Roma had gone back to work part-time as a community nurse in a local Aboriginal health service, which meant a gruelling routine of pre-dawn starts in the paddock to help round up the cows for milking before changing out of her dirty work clothes into something suitable for the school run. She would then drive Tom twenty-five kilometres to school in Purnim, turn around and take Austin to kindergarten, then go on to work. In the afternoon, she would do the whole operation in reverse, changing back into her milking clothes once the children were home.

'I used to drive almost 600 kilometres a week on the school run. I reckon my salary just about paid for the petrol.'

Joking aside (although what's funny about working for nothing?), Roma loved her job with the Aboriginal community. The only non-Aboriginal who wasn't a family member or married to someone in the community, she became an advocate as much as a nurse.

'When you're sick you're vulnerable and it helps to have someone who can walk you through the hospital system.'

Tom was self-reliant from an early age. When Roma came back from milking in the morning he would have Austin up, dressed and ready for kindy. Once his parents put a shocker on the toaster, the eight-year-old would also make breakfast and prepare all the lunchboxes. Resourceful and sensible, it was always salad in his lunchbox.

The farm was a great place for the children to play, especially Tom, who took after his quiet father with a shared love of the outdoors. Tom was nimble and agile, wrestling bull calves for fun and climbing trees to jump from branch to branch. His younger brother was less keen. Against his better nature, and urged on by his brother, Austin hauled himself up into a tree one day and came face to face with a koala. The shock was so great he fell out of the tree.

Tom's response was to climb up, sit beside the koala and make friends with it. When he noticed the docile creature had a baby he made friends with that too, and before long the female koala trusted him enough to let him take her baby out of her pouch, cuddle it then put it back. The relationship between Tom and the koala went on for months.

It sounded like they had finally found the rural idyll Roma had lost in childhood, yet they still didn't own their own farm.

It was Glenn who pushed them towards ownership. He may have been the introverted nature lover in the relationship but he was also a visionary and strategic thinker, reaching beyond where they were to where they one day could be.

'We're doing it again,' he said at the end of another long day. 'We're working our butts off and we're getting nothing back. Look at what we've achieved with this dairy we're managing. If we can do that for someone else, why can't we do it for ourselves?'

Talk in the industry at the time was all about deregulation and the removal of state and federal controls on the supply, distribution and pricing of milk. It worried dairy farmers in other parts of Australia far more than it did in Victoria. Research there had proved that deregulation would benefit consumers and Victorian dairy farmers had long since begun to prepare for its introduction.

Glenn and Roma were well placed to take advantage of any opportunities that might arise. In the past couple of years Glenn had carried on studying at night school and Roma had completed a diploma in agriculture. They had immersed themselves in dairy, consulted expert opinions and read everything they could get their hands on about the industry they had joined.

'Farmers are very knowledgeable people. We had a lot to learn and we found plenty of people willing to teach us,' said Roma.

They decided to make one last effort to buy their own farm. At the end of 1998 they found a small property near Woolsthorpe, three hours west of Melbourne, with 210 acres of flat volcanic soil, a herd of 100 cows and an outdated sandstone dairy. Roma prepared a budget, they approached the bank again and this time, after three years of study and a proven track record in dairy, the bank agreed to advance them half a million dollars.

It was the opportunity Glenn and Roma had been waiting for. Finally, they were able to set themselves up in business as farmers. They agreed a price on the property near Woolsthorpe, exchanged contracts and moved in on Mother's Day 2000, an auspicious date if ever there was one.

'And within five years we'll buy another farm,' Glenn confidently declared.

Many dairies at the time were using government money to upgrade to rotary platforms so Glenn and Roma were able to buy all the equipment they needed second-hand, aided by $30,000 of assistance from a fund set up to help with the adjustment to deregulation (money that came from a levy on farmers). They poured the concrete slab themselves, erected all the yards, and within six months they had built a new dairy.

'We went hell for leather. We decided that if we were going to borrow that much money we wanted to see a return on our investment as quickly as possible. It sounds like nothing compared to the amount we've borrowed since, but that was a lot for us in those days.'

It still sounded like a lot to me. How much had they borrowed since to keep their dream alive, I wondered? It was a question I wasn't game to ask.

While Glenn was welding steel for the holding yards, Roma would be milking, helped by twelve-year-old Tom, eight-year-old Austin and even five-year-old Vincent.

'They could all milk independently by the time they were ten.'

Part of the appeal of the particular farm they bought was the well-maintained house that went with it.

'We knew we wouldn't be able to invest in anything apart from the business.'

In doing all the work themselves, Glenn and Roma saved a fortune, spending $50,000 on a dairy that might otherwise have cost $200,000.

They forced themselves to ignore doomsayers prophesying the end of the road for small dairy farmers, a gloomy outlook shared by many in the Australian rural press, but it must have been hard to stay positive when dairy farmers in New South Wales, Queensland and Western Australia were struggling to cope. The dire predictions even reached as far as England, as this extract from the *Guardian* newspaper on 1 November 2000 demonstrated:

> Deregulation of the Australian dairy industry has resulted in thousands of smaller farms going to the wall as their farm gate price for drinking milk plummeted from around 56 cents to 27 cents.

Roma had no truck with the notion that there was no sustainable future in agriculture. 'Even members of our own family were telling us we were stupid to be joining the dairy industry, but we knew what we were up against and we budgeted accordingly.

'We arrived in May, finished building the dairy on 1 November, and milked 200 cows that day.'

The novice dairy farmers doubled the size of their herd in the first six months.

As Roma described their ambitious plans and the intense pressure of those first few months, I couldn't help wondering if they'd bitten off more than they could chew. Maybe waiting so long to achieve their dream had unintended consequences? Did they sprint for the finish line as soon as the gun went off instead of easing themselves into a long-distance race?

'Life is ten per cent what happens to you and ninety per cent what you do about it,' said Roma. It was a piece of advice she frequently used to bolster everyone's confidence (including, I suspect, her own).

Another was, 'Achieving anything in life is tough, if it was easy everybody would be doing it.'

There was nothing easy about their early days in business for themselves. Roma came off a decent nurse's wage to work in the dairy and their first milking cheque was for $5000. Their bills were $10,000.

With no let-up in work everyone was expected to pitch in, including all the children, who collected eggs from the chooks, fed the pigs, pulled up vegetables and regularly helped their parents in the dairy. The huge willow tree that grew beside the holding yard for the cows became their playground, even for the reluctant Austin.

One of their favourite playmates was a pet possum, tame enough to submit whenever they wanted to play with it. During milking they would carefully place the possum in the tree, and when they'd finished playing with it they would put it in a cage overnight.

One evening when milking had finished the children searched in vain for the possum. When they couldn't find it they ran up to the house, worried it might have met with an accident. Roma was able to reassure her anxious children. Earlier that evening, standing at the kitchen window, she'd watched the possum climb onto the back of their Jack Russell, Ernie, and hitch a ride back to the house. The possum was now safely back in its cage.

Tom even took a baby possum to school with him once, hiding it in his coat pocket and surreptitiously feeding it with a dropper during lessons. When the less discreet Vincent tried to emulate his big brother the school called Wildlife Rescue.

With so much pressure during the week, Roma and Glenn made a conscious effort to prioritise family time at weekends.

'We made sure we had someone working with us at the weekend, even if it was only for one milking, to give us all a bit of time off.'

Two years after they went into business, in August 2002, Roma discovered she was pregnant again.

'I was thirty-six years old, the same age my mother was when she had me.'

Roma had often referred to the parallels between their lives which thankfully didn't include the tragic loss Pauline suffered. Her mother had been a strong-minded career woman and Roma clearly took after her.

Having always thought she would have four children, Roma was delighted at the news. She expected her fifteen-year-old son Tom to react with a shudder of revulsion at the thought of his mum having another baby but he surprised her.

'He was thrilled. He asked if the baby could sleep in his room.'

I had no problem believing Roma when she told me earlier in the day that she was chopping wood an hour before Austin was born. I was less inclined to believe her when she said she put her feet up for her final pregnancy. She assured me it was true.

'It was the boys,' she explained. 'I had a wobble in a paddock one day and it was probably just from drinking too much caffeine but from then on they insisted I rest up.'

Accustomed to hard physical work in her previous three pregnancies, Roma was surprised to find she was treated with kid gloves.

Tessa Monique was born in April 2003, amidst signs of a reprieve in the prolonged drought that had begun in parts of Queensland in 2001 and gradually spread across much of western New South Wales. The drought had impacted on eastern South Australia, the Gascoyne region of Western Australia and north-western Victoria. As the decade advanced, so did the drought.

Notwithstanding the arrival of another child, or the drought, Glenn was convinced they should leverage off their debt to expand. The international demand for milk was growing and land prices had increased in value over the past few years.

Neither he nor Roma were interested in building up a business so they could retire wealthy; their plan was to build up an asset as an investment in their family's long-term future.

'I started with nothing and I'm going to go with nothing,' Glenn said, making me wonder again at the size of their debt.

They had been leasing a block of 110 acres for some time, which had allowed them to increase the size of their herd, so they approached the owners to see if they had any appetite to sell. There was no interest so they shopped around and found a farm within a few kilometres of their own. The bank agreed to advance them a mortgage and Roma and Glenn put on a manager to run the new dairy.

Six months later, sitting at the kitchen table with the local paper, they turned automatically to the property pages. There was the farm they had tried to buy, the one where they were leasing a block of land. The owners had clearly had a change of heart.

'We read each other's minds and we both knew we were going to buy it,' said Roma.

There was another trip to the bank, a bigger mortgage and yet more managers employed on the new farm. In six years Roma and Glenn had gone from milking a hundred cows to milking a thousand. It was a haphazard, opportunistic growth spurt that plunged them into debt and left them struggling to cope.

Life got very busy and it grew increasingly difficult.

*

Dinner that night was like a gathering of the United Nations. There was the breezily confident Yuri from an eastern European farm in the Ukraine, big tall Nelson who had recently arrived from Ecuador and was still finding his way around, and shy skinny Alex from Yorkshire in northern England, who had never milked a cow before she'd arrived and was now on her second three-month stint at the Britnell's dairy.

The appeal for all the young foreign workers joining us for dinner was life in a family environment on an Australian farm, with up to twelve months' paid work at the normal Australian wage. Some stayed for three months; others, like Isalaene from South Africa, were more permanent. For Glenn and Roma, employing a series of workers on an international exchange program had eased the pressure of running three farms.

It was a cold wet night and the wind had picked up, flinging rain against the uncurtained windows as Tessa carefully laid eight places at the table her father had built. It turned out Glenn built most of the furniture in the house.

'You have to wait ten years but you get it in the end,' said Roma, pulling a quiche out of the oven.

Judging by the quality, it was worth the wait.

Isalaene appeared towards the end of dinner, delayed by a Zumba class she'd attended. The experienced dairy hand from South Africa saw to her sixteen-week-old whippet before joining the table.

'Are you glad you didn't go to Zumba?' Isalaene asked Roma,

her deep voice carrying a strong Afrikaans accent, as she reached for the last piece of quiche.

'I'm *always* glad I don't go to Zumba. Who was there?'

'A man,' Isalaene said in a measured way.

'Single?'

Isalaene chuckled quietly, knowing Roma was trying to set her up.

'You should have been there. I don't want you to be jealous when I wear a bikini in summer.'

Both women laughed at the suggestion that either of them might be wearing a bikini anytime soon.

Roma clearly enjoyed the company of another adult female, especially one who knew as much about the dairy industry as Isalaene. She had been with them for two years and her help in that time had proved invaluable. Isalaene was another quiet introvert, like Glenn, and she knew a lot about dairy farming.

'We grew so fast it's only been in the past two years that we've finally got the right systems in place,' said Roma. 'Getting good people matters but you need systems and procedures to support them.'

'I'd say it was only in the last twelve months,' Glenn countered quietly.

The discussion turned to calving rates and tractor repairs, killer whales that had been spotted off the coast near Bells Beach and the merits of growing garlic.

'Isn't it time for you to feed that lamb, Tess?'

'In a minute.'

Tessa was enjoying taking part in the discussions around the table. Being the youngest, she sometimes got more of her mum's attention, and sometimes less.

'I always used to make it to every sports day but I think I've missed two of the three medals Tess won.'

'Mum, I've got four now!' Tessa piped up, carrying dirty dishes over to the sink.

'Have you?'

'Yes, one for hockey, one for the under-nine champion, one for school sports . . . and . . . and . . .' Her face fell. 'What was the other one for, Mum?'

'I didn't miss *that* one.' Roma laughed.

'Oh yes, I remember now!'

Tessa had been competing in the 400-metre sprint when she fell over halfway down the track, badly grazing her elbows and knees. She got up, overtook the other competitors and crossed the finish line (crying) in first position.

'That medal should have blood on it!' said Roma, making the excitable Tess giggle.

'What do you want to do when you grow up, Tessa?' I asked.

'In between a vet and something to do with animals, only not dairy because I don't like the smell and it's a dirty job.'

This was a girl who knew her own mind.

'It's like a family thing, Tom or Austin will take over the farm and Vincent will become a lawyer and I will become an animal nutritionist maybe.'

'Why can't you take over the farm?' Roma asked.

'Because they're the oldest.'

'If you didn't have brothers would you take over the farm?'

'Oh yes.'

'Well, I suspect it will be you or Austin who takes over,' Roma said, getting up from the table.

'Well, Austin likes animals, he just doesn't know anything about them.'

Everyone laughed at Tessa's forthright opinion as we handed over empty plates, tipping scraps into a bucket for the pigs.

'The great thing is it comes back to us as bacon,' Roma said.

'So we eat the mould and the scraps again? I find that disgusting,' said Tess.

'Yes, Princess, you would. It's called recycling.'

'It's called crackling,' said Glenn, whose occasional remark in the hubbub of loud conversation was always perfectly timed.

The young helpers disappeared to their own space and we finished the evening in the lounge room, where Roma offered everyone a hot drink then promptly forgot what Glenn and Isalaene wanted. Glenn was clearly used to it.

'You're too busy to listen to yourself,' he said, switching on the kettle again.

I wondered if the rapid expansion of their fledgling business had tested their relationship, especially with two such extreme personalities.

'There used to be a lot of conflict,' Roma admitted, likening the way they worked to being in a ward with five patients. 'We were both looking after the same patient. One of us would come in and wash one arm and five minutes later the other person would come in and wash the same arm again.'

Their problems were exacerbated in 2005, when drought returned with a vengeance.

Harsh as it was for those forced to sell, it was also an opportunity for Glenn and Roma to buy cattle cheaply from the north and restock their dairy. Knowing the thousand-kilometre journey would have an impact on the cows, they budgeted for milk production to drop from twenty-five litres a day to eighteen.

The night the cows left northern Victoria for the final leg of the journey it was a comfortable twenty-eight degrees. Then the weather changed.

'We suffered the only thunderstorm this country saw that year.'

The cattle arrived in the middle of a hailstorm and temperatures plummeted to twelve degrees. The dairy cows contracted pneumonia then salmonella and some of them died; those that survived struggled to produce a meagre eight litres of milk a day. At the same time, Victoria ran out of hay and the Britnells were forced to truck feed in from Western Australia. Instead of costing $200 a tonne it cost $500, and now they had a thousand cattle to feed.

When the cows started dying, not for lack of feed but because of an obscure fungal infection, the situation looked dire. On one day alone, twenty cattle died over a six-hour period.

They traced the source of the infection to pasture in a particular paddock and instructed the manager on the Port Fairy farm not to use that paddock. He ignored their advice, put the cows back in and more died.

It was the final straw for the embattled Britnells and worse was to come when their attempts to sack the manager led to a protracted court case.

'It was horrifically hard,' said Roma.

The one bright spot in all the chaos was a Welsh couple managing their third farm, reliable people with good experience of dairy cattle, who valued honesty, respect, care and hard work. Their children went to the same school as Vincent and their steady support gave Roma and Glenn a glimmer of hope.

'We learnt the importance of employing solid people with values that were similar to our own.' Put simply, they looked for – and found – managers who were good people.

There were casualties though. Their frenetic period of growth coincided with Tom's teenage years, from the age of twelve to seventeen, and Glenn and Roma both agreed that the intense pressure of those years probably put their oldest son off farming.

'It was chaotic, we didn't have structures in place and Tom worked incredibly hard.'

Tom had always said that he wanted to go to university but as the time approached he seemed more interested in pursuing a relaxed approach to life than in doing the kind of hard study university would require. Roma was on his case straightaway.

'You can leave school or you can study, but you have to choose and then you have to follow through.'

Tom's young girlfriend, Hayley, flinched at such open conflict and it can't have come as much of a surprise when Tom decided to leave school after year twelve. It wasn't long before he left home as well, and he moved away from farming altogether.

Seeing how hard Tom had worked Glenn and Roma knew they would have to get help if their second son, Austin, was to find enough time to concentrate on his studies. Austin had always talked of becoming an entomologist, even when he was five and didn't know what an entomologist did.

By the time Austin reached year twelve the Britnells had put on full-time milkers. Vincent was even luckier, he stopped milking in year nine, 'and Princess has never milked,' said Roma.

Throughout their children's formative years Glenn and Roma did everything they could to prioritise family time. They were at the finish line for every school sports day and on the touchline for every footy game; they took at least one milking off every weekend and, no matter how busy they got, they always had a ten-day family holiday every year. Camping by a muddy lake forty kilometres away from home may not have been the height of luxury, but surrounded by other families they knew it was always a lot of fun.

'We have the best memories of those holidays,' said Roma.

To Roma and Glenn's relief, Tom didn't stay away from the family or from farming for long. After a stint as a mechanic he went on to study agronomy and, two years ago, the 25-year-old moved back to manage their Port Fairy farm, with Hayley, who was now his wife, by his side. Hayley worked as a dental nurse in nearby Warrnambool and she no longer recoiled at some of the things her forthright mother-in-law said. In turn, Roma had wholeheartedly embraced her daughter-in-law.

'She's a fantastic girl, the best daughter-in-law I could ever have hoped for. Tom chose well,' she added with a smile.

All three oldest children were living away from home when I visited; Tom at Port Fairy, Austin at university studying agricultural science with a view to majoring in entomology or business, and Vincent at boarding school, where he had developed an interest in political science and debating.

'He's won just about every debate in the state for the last six years' said Roma, proudly.

Vinnie, as they called their third son, had also inherited a love of animals and a taste for travel. At fifteen he spent the summer in Europe visiting a backpacker who had once worked with Glenn and Roma in Victoria. After that he worked at a robotic dairy in Denmark followed by a six-week stint in a Danish zoo, owned by another ex-worker.

'He came back and he was suddenly a grown-up,' said Roma, a little wistfully.

Roma seemed ideally suited to the challenge of running a business with a multi million-dollar turnover yet for a long time she resisted getting too closely involved, insisting Glenn was the farmer, not her.

In the end, they took a long hard look at their strengths and weaknesses and Roma took on the role of business operations manager while Glenn concentrated on looking after the farming side of things and the skilled maintenance.

I was impressed at Roma's insight into her own character.

'I'm a better communicator and people person but if it needs a quieter, calmer approach Glenn takes over.'

The changes they implemented worked well and life had finally settled down into a semblance of normality . . . or what passed for normal in the Britnell household.

Roma retired from her part-time job in Aboriginal health in 2012, after fifteen years working in a field she loved and forging close friendships with many people in the community. She was proud of the fact that the community had built its own health centre.

'My aim was always to make myself redundant.'

Giving back to the industry that gave them a start when everyone else knocked them back had been important to them both. The resources of United Dairyfarmers of Victoria (UDV) were vital in helping them make wise business decisions in the early days of working in what was then an unfamiliar industry, and they both still attended regular meetings.

Roma was an especially passionate advocate for the industry. UDV sent her on a leadership course in Canberra when she was pregnant with Tessa and she repaid their investment by joining the board of West Vic Dairy. She was also a policy advisor for UDV.

The achievements kept coming. In 2009 Roma was awarded Australian Rural Woman of the Year, she won a Nuffield scholarship in 2011, she was appointed to the Victorian agriculture minister's Women in Primary Industries Roundtable in 2013, and in 2014 she was elected vice-president of UDV.

It all made for a busy life.

The following day, Roma would be up before dawn to deal with paperwork at the kitchen table before rousing Tessa and getting her to school. Later there would be volunteering duties in the school's kitchen garden program (she was also on the board of a new school being built in Woolsthorpe) then she would be food shopping for the young backpackers, ringing farmers and writing policy documents. There was a meeting of West Vic Dairy to attend in the evening, a gymnastics class for Tessa and a long drive home in the dark several hours later.

'Right, does everyone know what they're doing tomorrow?'

Roma ran through a list of jobs, making sure Nelson knew he had to register his licence so he could drive the car and suggesting Yuri might like to drive him and Alex to the coast the following day to look for whales. She was simultaneously logging on to her computer to check the arrival details of a new girl from Canada.

'She arrives Sunday, did I read that in an email? Hang on, we're going to Hall's Gap on Monday . . .'

Glenn smiled. 'Our society is geared towards extroverts. They are the ones that people always listen to because they are the noisy ones.'

He looked at Roma as he spoke and she laughed. Their relationship was a good one, strong enough to withstand robust debate and loving enough to incorporate gentle ribbing. She turned her attention to Glenn and asked his opinion on a problem with the surge pulsator in one of the dairies.

Roma Britnell may have been the extrovert in the partnership but she long ago learnt the value of listening to introverts.

Ian and Merry Jackson

'Tirlta'
115 kilometres northeast of Broken Hill, New South Wales

The remote mining town of Broken Hill in far northwest New South Wales is surrounded by sandy desert and low-lying scrub; out there, the bush is everyone's backyard. I had lived in Broken Hill when I worked for the Royal Flying Doctor Service, and I'd met a number of station owners, so I was keen to include a local grazier in this book.

I found Ian and Merry Jackson through Joe, a Broken Hill man whose natural affinity with the bush led to a lifelong friendship with the Jackson family. Decades back the two families helped start a Broken Hill chapter of the American Field Service (AFS), and for the next twenty-odd years Joe and his wife Marlene helped organise annual events at Ian and Merry's station, Tirlta, welcoming wide-eyed students from around the world and introducing them to the unique beauty of the Australian Outback.

Tirlta sits on a patch of sweet country near Mutawintji National Park, about an hour and a half's drive from Broken Hill. Drive out of town in any direction, to Cobar, Menindee, Adelaide or Tibooburra, and it's hard to believe there are station properties hidden behind the low-lying scrub that stretches as far as the eye can see. The only indication might be a battered mailbox, a faded sign or an old fridge where a sandy track veers off the bitumen and disappears into the distance, with no indication of how far you might have to travel before you reach your destination. The track that led to Tirlta ended at an oasis of beauty in an otherwise arid landscape.

Ian had a pocketful of stories to tell about his family's history, and, in that great Aussie tradition of bush balladeers spinning a yarn, most of them played up the larrikin side of human nature. At times, Ian stretched my credulity to breaking point but his delight in sharing his surprising tales, his passionate commitment to family and his eagerness to celebrate their achievements always pulled me across that scuffed line in the sand between fact and fiction. And at the end of a long hot day in the bush, who cares how much embellishment might have been added over the years? Storytelling has always benefited from a touch of dramatic licence.

Ian's wife, Merry, was as quiet and serene as her husband was outgoing and gregarious. She would occasionally interject with a gentle smile and I sensed she would have pulled Ian up if he had strayed too far from the truth.

The visit I made to Tirlta turned out to be one of the most entertaining I undertook in researching this book. While I was staying with them, Ian and Merry drove me out to their ephemeral lake at sunset, where we met up with one of their sons, Matt, his wife Sarah and their two children, Sam and Archie.

As the sun disappeared on that hot spring day we took part in a game that even now I struggle to explain. It involved four of us lifting a seated adult high into the air using only our fingertips. It was a mind-over-matter game that took place against the backdrop of the swollen lake, ink-black under the night sky.

I don't recall much apart from Ian's eyes, sparkling in the darkness. 'Now you know my secret,' he whispered.

Ian and Merry shared many stories with me during that visit, but none were quite so astonishing as that mystical show of force on the banks of an ephemeral lake near Broken Hill.

He's no dumb bastard

'I hope we can make a contribution to your book. Do you think we're interesting enough?'

We'd been talking all day. It was a warm spring night and a faint glow from the distant town of Broken Hill lightened the dark horizon; the Barrier Ranges were out there somewhere. Unseen from our vantage point on the upstairs balcony, hidden in the dark shadows, lay the remains of a humpy Ian's grandfather had lived in, the shearing shed he'd built, the shed his father had built, and the gleaming new woolshed Ian had built. There were wells and bores hundreds of feet deep that his forebears had sunk through silt by hand, a tennis court and swimming pool his father had put in, and a museum housing the most remarkable collection of Australian memorabilia I'd ever seen.

Within hours of meeting Ian Jackson I'd gathered enough stories to fill a book, never mind a chapter. He'd shaken hands with a movie star, an Olympic athlete and the most notorious American president of recent times – all before the age of eighteen. He'd risen from dunce to dux at school, survived a near-fatal car crash only to die briefly on the operating table, and he'd gone from borrowing money to feed his first child to owning not one but two thriving sheep and cattle stations.

Was he being disingenuous? Ian Jackson played by no rules but his own and he could be charming, bullish, forthright, modest and supremely self-confident in the space of a single sentence. Like an old-fashioned showman, his stories were sprinkled with a touch of magic, and his eagerness and enthusiasm to share them was infectious enough to suspend disbelief at some of the more outlandish claims. He made most people's lives (mine included) seem tame by comparison. Somehow I didn't think he was fishing for a compliment when he posed that remarkable question. Family meant everything to Ian and he genuinely wanted to know if I was interested.

'Yes,' I said. 'I think your story is interesting enough to be included.'

He swelled with pride and pleasure. Looking back on photographs I took on that trip, I realise now that Ian's larger-than-life personality made him seem bigger than he was; in fact he was only three or four inches taller than his slender wife.

'If you decide to write about us, just let it flow,' he said.

Merry's laughter in response to such a peremptory command sounded like a bell chiming in the night.

'And don't make it sound proper,' he continued in his gruff voice. 'I'm not proper so just say it as it is. Don't make it better. Make it natural.'

I might have bristled at such orders from anyone else but that night I couldn't help agreeing with Merry when she said, 'He's such a dear.'

And he was.

So here goes Ian (and I'm sorry about the odd long word, I didn't believe Ian for a nanosecond when he said he was the dumbest kid in school).

Ian and Merry were effusive in their welcome and admonished me for arriving with supplies. I knew I'd done the right thing though, having finally worked out that you shouldn't arrive at any station empty-handed. That didn't mean the wine and chocolates I'd offered my hosts elsewhere, it meant fresh bread, milk and the morning papers – the kind of luxuries we take for granted in city suburbs with corner shops.

I had driven to Tirlta from Broken Hill, following the Silver City Highway to Tibooburra for a hundred kilometres or so, watching for signs to the entrance to Mutawintji National Park. It was a hot day in late September, heading for a top of thirty-six degrees, and a skittish wind kicked up puffs of red dust under the clear sky.

The turn-off would have been easy to miss without clear directions. As I bumped along the four-mile sandy track that led to Tirlta there were no signs of the thousands of sheep and cattle I knew were out there somewhere; all I passed was the occasional fat lizard basking in the heat and a surprisingly dense cover of scrub.

Within minutes of arrival I was ensconced in the best room in the house, a handsome bar-cum-lounge room upstairs. A second storey was a rarity for a station homestead – even most houses in Broken Hill were single storey – and it was a treat to get a different perspective on the largely flat bushland beyond.

'We had a big flood in 2000. The house fell apart so we rebuilt it,' said Ian, settling himself onto the sofa.

Merry appeared up the stairs with a jug of cold lemonade and a plate of biscuits. 'Ian did a lot of the building himself.'

'We had all the gear. Just built it when we had a spare moment. It was easy.'

Easy to build a house? (Try telling that to anyone who's battled builders in Sydney.) And suddenly we were off, gathering speed as Ian rattled off details of his extraordinary family history, enlivened by philosophical musings and occasional brief pauses to gauge my reaction.

My raised eyebrows only spurred him on.

Ian's grandfather, Aubert Mussen Jackson, came up from

Melbourne 'way back in the 1800s' and found work on a station near William Creek, an isolated township on the Oodnadatta Track in northern South Australia.

Aubert moved around, working from station to station, until he met his wife, Ida Sparks, whose father was a prominent Adelaide businessman.

'He emigrated from South America, bought a 240-tonne frigate in Argentina and settled in South Australia.'

Henry Sparks found fame as a prize-winning student and a notable sportsman. He went on to serve as a justice of the peace; on the synod of Christ Church, North Adelaide; and as mayor of Glenelg council. (All true, I checked.)

Ida and Aubert settled first at Tarcoola Station then Kara near Broken Hill, where Aubert was known to be a hard worker. He cut the stone out of the hillside at Kara to build the family home and he took work wherever he could find it, including at Langawirra.

'Grandfather's father somehow had a share in Langawirra, no idea how . . .'

Nor me, but never mind. Ian's grandfather had been fascinated by Aboriginal history and he frequently traded dry goods for spears and artefacts, building up an important collection that Ian still treasured today. The promise of seeing some of those artefacts was one of the reasons why I'd been keen to visit.

'So then Kidman took over Yancannia in 1916. He asked Aubert to work there and RM Williams was there at the same time and he taught Dad to plait cattle hide.'

It was a struggle to keep up as Ian's storytelling gathered pace. Unexpected switches of direction felt like a horse's tail swatting at flies. I hoped my trusty tape recorder would capture enough to make sense of the story later.

Aubert jumped at the chance to work on Yancannia (one of

the largest stations in New South Wales, spanning 1487 square miles west of the Darling River) and he moved there with his wife and their three children – Harry, James and John.

The prosperity that followed was coupled with tragedy. Aubert and Ida's youngest son, John Jackson, never fully recovered from a bout of meningitis that left him needing constant care, and in 1934, 55-year-old Aubert suffered a fatal heart attack while umpiring a cricket match at Wonnaminta Station. His oldest son, 23-year-old Harry – Ian's father – had been bowling.

Following the death of their father, the two younger sons, James and John, stayed at Kara and Harry was offered Aubert's share in Langawirra. Harry discovered which end of the station he would inherit through the simple act of drawing straws.

The partner who drew the long straw kept the area near the lake, with all the established buildings including the homestead, and the partner who drew the short straw got the rest of the property.

'They drew lots and Dad lost.'

EP Tapp, who had been a partner with Aubert, consoled Harry. 'Don't worry, laddy,' he said. 'You missed out on the station but you got the best country.'

It was small consolation to young Harry, who was forced to live in a humpy he built himself out of bits of corrugated iron.

'He didn't have a cracker,' said Ian, referring to his father's lack of funds.

Once the estate was settled, half the shearers' quarters, half the yards and half the stables were pulled down and Harry erected them near his humpy. Several months later he discovered he'd built them in an area prone to flooding. When the first big rains came the sheds were inundated and the whole lot had to be pulled down and moved again.

Harry Jackson sounded like a resourceful young man, and if only half the stories Ian told me were true, he was also a wild larrikin with an unbridled sense of fun.

West Darling Week was an annual event that closed the main street of Broken Hill for several days of raucous celebrations and wild goings-on, an ideal opportunity for Harry to dream up ever more fantastical practical jokes. He once dressed a ram in ribbons and took it into town for the event, tying it to the leg of a table in the dining room of the Royal Exchange Hotel while he and his mates had dinner. When a female guest complained, he hid the ram in the ladies toilet, waiting with quiet pleasure to hear the horrified screams he knew would follow. When he was forced to remove the ram, he rang a taxi.

'I'll give you ten pounds if you'll take a passenger anywhere in town.'

'You're on!' the delighted taxi driver said; ten pounds was a lot of money in those days. He was less pleased when he discovered the passenger wasn't human.

'I want you to take this ram to the police superintendent's house and put it in the garden.'

'In the garden?'

'Yes, he specifically said he wanted it in the garden. And don't tell anyone, it's a surprise.'

It certainly was. The police superintendent woke the next morning to discover his entire collection of prized roses had been eaten.

Ian laughed. 'And they never knew who did it!'

They do now, Ian.

Harry met his wife, Esther, at West Darling Week and she was as game for fun as he was. Sent to a Presbyterian boarding school in Adelaide at the age of five, Esther grew into a fun-loving young woman who would tuck her dress into her bloomers to race unimpeded along Argent Street during West Darling Week.

'Mum was a goer and she was also kind, gentle and gracious. She never said a bad word about anyone.' Well, apart from the time during World War II when she climbed onto the rotunda in Burra – a town with a not inconsiderable number of German settlers – to publicly declare war on Germany.

It struck me Ian had met a similarly kind, gentle and gracious person in Merry. She sounded a lot like his late mother and I wondered how two such different people had come together.

But there was a lot of talking to do before we reached that part of Ian's story.

We took a break on the sheltered balcony where house martins swooped frenetically under the eaves, urgently building mud nests as fast as Merry and Ian could knock them down.

'Little bastards,' Ian muttered.

'It feels cruel but they make such a mess,' Merry explained.

The spars and paddles of an old windmill stretched across the sparse lawn and Ian pointed out the tennis court his father had put in sixty years before – 'all the neighbours used to come every fortnight for a game' – then the swimming pool his father had dug. Nearby was an old-fashioned plough, rusty and disused now.

'The old man was only eight stone and that's the plough he used to cut the dirt up.'

Ian saved everything that had an attachment to family history.

Beyond the fence, bare patches of red earth were far more plentiful than the sparse trees and thorny bushes; inside the fence line, flowering bushes and mature fruit trees flourished. I spotted orange, lime and lemon and Ian showed me a stand of quandong trees that had been fenced off to stop feral goats from eating them.

'We had some kids out from Melbourne, took them to Mutawintji and showed them a quandong tree. They had to write a report when they got back and the teacher put a red line through the essay: "You don't spell condom like that and Aboriginals don't eat them."' Ian shook his head in disbelief. 'I showed this girl the exact spelling and she got penalised.'

Ian showed me one of the tiny bush peaches later, peeling away the mottled skin to reveal a thin layer of flesh that tasted sharp and intense (perfect for jam).

From our vantage point, looking out across the seemingly empty windswept land, it was hard to believe Ian's stories of vast herds of feral goats. His brother had trucked out 13,000 in less than two weeks, not counting the 9000 he let go because they were too small.

'Some goats are under wire but a lot of the ones you see around here are wild.'

There wasn't a single wild animal to be seen. I was reminded yet again of just how vast the Australian Outback was.

Ian was one of four boys, two older and one younger, and his grip on life was tenuous at first. Born with polio, which left him with a crippled leg needing multiple operations, Ian survived scarlet fever, appendicitis, a bout of rheumatic fever that almost killed him and glandular fever.

Given what a larrikin Harry Jackson had been, he must have assumed that at least one of his sons would inherit a wild streak. Notwithstanding his poor health, Ian vied with his brothers for that inheritance.

'We were the wildest blokes you ever put breath in.'

Ian's partner in crime was generally Pete, a year older and up for anything. When they were still too young to attend school the two brothers proved impossible to control, so one day their exasperated mother resorted to drastic measures. Knowing the two were inseparable, she tied one of them to a tree. As Esther had suspected, the unfettered child played happily in the dirt beside the one tethered at the end of a twenty-foot rope. From then on, they took it in turns to be tied up.

Does that sound shocking? It's not really that different to the reins I wore as a child whenever we went out. Or the playpen I spent hours in. And on a property of 96,000 acres – with dams, wells, bores and a lake – how else could Esther control four 'feral' boys?

They never wore shoes, they rarely wore clothes and, as soon as they reached school age, the peal of the school bell each morning was their signal to run and hide. When one of their exasperated parents eventually found them the boys would stomp into school with slug guns hidden in their pockets. As soon as their long-suffering governess turned her back they would take their guns out and shoot blowflies crawling up the wall.

The boys would often climb a tree to avoid being dragged into the schoolroom and it was only hunger that eventually lured them down. The flogging that followed made no difference.

'For some reason that just spurred us on.'

The night they dumped their governess was extreme though, even by their standards.

*

'Where's Vivien?'

'Don't know, Dad.'

'Well go and find her and tell her dinner's ready.'

'Okay.'

Seven-year-old Ian and eight-year-old Pete ran out of the kitchen.

'Vivien! Dinner's ready!' They made a great show of calling their governess's name, banging their fists on her bedroom door. When there was no reply they dashed back to the kitchen.

'She's not there, Dad.'

'Check the schoolroom.'

Ian and Pete ran across the dusty yard and stood in the dark outside the old sinker's hut that served as a schoolroom.

Ian shouted at the top of his voice. 'Vivien! Where are you?' His muffled laugh prompted Pete to punch him and the two scuffled in the dirt until their father's appearance broke up the mock fight. The big man loomed out of the darkness and towered over the boys.

'Right, you buggers, what have you done with her?'

'What d'you mean, Dad?'

'You know exactly what I mean and if you don't tell me where Vivien is right now I'll get the whip out!'

Harry Jackson was a softly spoken man who wasn't given to swearing or blaspheming but there were times when his four boys tried his patience beyond breaking point. When that happened the riding crop came out. It had already been worn down to half its length and it looked like it would be needed again that night.

Knowing they were cornered, Ian and Pete confessed. Earlier that evening they'd taken their governess roo-shooting, bumping along miles of sandy tracks in a cut-down Volkswagen that they'd both been able to drive since the age of seven (and it's only now as I write that sentence that I remember I wasn't shocked at them having guns; I was shocked at them driving).

It was getting dark when the boys parked the car, telling Vivien it was her turn to shoot. As soon as she stepped out of the car they gunned the engine and took off in a cloud of sand, leaving Vivien stranded on an isolated dirt track six miles from home.

It was totally dark by the time they started searching, and they searched for hours, finally locating the sobbing governess in the early hours of the morning, near the ten-mile swamp. I'm not one for corporal punishment but I think even I would have flogged the boys that night.

Astonishingly, Vivien didn't resign her position. Ian thought the reason for that was simple. 'She wanted to exact revenge on the four young tearaways who regularly filled her pyjamas with cannonball prickles.'

I trust she got it.

Like his two brothers before him and his younger brother after him, Ian was sent to St Peter's College in Adelaide, a six-hour drive from Tirlta. In his first school report, at the age of nine, he came bottom of the class.

'I was the dumbest one there out of 1300 kids.'

Ian's defiant look challenged me to say different.

By his mid-teens Ian had grown into a robust lad whose aspirations centred on jackarooing, shooting roos, riding a motorbike and playing footy – he certainly didn't want to spend any time at school. He and his cousin, Andy Kerr from Yancannia, took it in turns to come bottom of their respective classes.

On the advice of one of his older brothers, Murray, Ian gave up smoking. 'We grew tobacco and we smoked tea-leaves but Murray said because I'd been sick it might kill me.' Ian was thirteen when he kicked his habit.

The one subject Ian did enjoy was geography. By March 1967, when he was fifteen, Ian had done well enough in his favourite subject to drag himself out of the bottom set. In a lull after exams, the geography teacher showed the class a form he'd been sent detailing an American scholarship scheme. Ian liked the sound of it and told his geography teacher he was going to apply. The response was less than encouraging.

'What hope have you got, you dumb bastard?'

(Can you *imagine* that today?)

The response prickled Ian into writing to his father.

Mr H Jackson
Tirlta Station
c/o Broken Hill Post Office
Dear Dad
Please can you send me $2? The geography teacher said
there is a scheme that sends kids to America. If I get chosen
I get to live in America for a year and I get to go to school
there. It costs $2 to apply. Thanks Dad.
Ian.
PS School is ok. I've gone up to grade C in geography now.
Ian Jackson
St Peter's College, Hackney Road, Adelaide

Dear Ian
I read your letter with interest and I wish you all the best with
your application to study in America. Please find enclosed
the money you requested. I believe this is the easiest $2 I
have ever wasted.
Dad
PS Your mother and I are both well and she sends her love.

More than forty-five years later Ian could still laugh at the
memory of that letter. 'What surprised me was that he enclosed
the money!'

Spurred on by suggestions he'd never make it, Ian powered
through the selection process for the American Field Service
program, writing essays, taking part in debates and joining a
family dinner to prove he could handle himself in a social setting.
Of 186 students who applied from South Australia only sixteen
were chosen. Ian was one of them. The final hurdle was waiting
to see if a suitable family could be found to accommodate his
wish to live in 'John Wayne country'.

Sam Benedict's family welcomed Ian to Arizona in the summer of 1968 by declaring, 'You, son, are a VIP.'

The surprised sixteen-year-old didn't know what a VIP was, and when he did find out he was even more surprised to learn they considered him one.

'This town raised a lot of money to have you come and stay here so we want you to dig in and do your best.'

The Benedict family treated Ian as one of their own, encouraging him to call them Mum and Dad and bolstering his self-confidence, without ever letting him slacken in his studies. Ian credited the experience and the environment with teaching him the value of hard work.

'I graduated with a straight A average.'

It was an unforgettable year in which Ian got to meet his hero John Wayne at the movie star's ranch, where they talked about cattle and his life in pictures. 'He was a big, gentle man, tall enough to rest his arm on my shoulder.'

At school Ian was coached in athletics by a world record-holder who went on to compete in a total of four Olympic Games. In 1968 George Young won a bronze medal in Mexico City, despite 'hitting the wall' 300 metres from the finish.

'He picked himself up and forced himself to keep going. I learnt a lot from him.'

Another defining moment was 21 July 1969, when Ian found himself in New York's Times Square as Neil Armstrong took his historic first steps on the moon.

'People were screaming and shouting, the whole place erupted.'

And one of the last events of his twelve-month stay was a trip to Washington DC, where 3000 AFS exchange students from around the world had been bussed in from homestays across America to gather on the lawn in front of the White House.

Ian got off a numbered bus and joined a throng of students chattering in dozens of different languages, all searching for their national flag to indicate where they should stand. The Australian flag was right at the front, a few metres from the rotunda.

A great cheer went up as the recently inaugurated thirty-seventh President of the United States of America walked onto the stage. Richard Nixon waved at the crowds and Ian had a sudden flash of memory, recalling the time his father had gone to a concert and climbed on stage to dance with the singer, Sabrina.

On impulse Ian turned to the girl standing next to him.

'I'm going to meet the president,' he said.

She laughed. 'You can't do that.'

'Watch me.'

Nixon finished his speech, waved once more to the crowd and Ian grabbed his chance. He strode to the edge of the podium and scaled the rotunda like an agile monkey. Startled bodyguards rushed forward, but not in time to stop Ian sticking out his hand.

'G'day, Mr President. My name's Ian Jackson and I'm from Austray-lia.'

A great cheer went up from the crowd and Nixon waved his bodyguards away. He held out a hand that Ian remembered as pale and spongy.

'Hello young man,' Nixon said. 'It's a pleasure to meet you.'

Ian took me on a quick tour of the property later that day, pointing out a huge coolroom that his father had built fifty years before – another 'first' for its time. The sign along the top looked Aboriginal so I looked closer – 'didyabringyagrogalong'. Sure enough there were plenty of cold beers inside stacked on battered metal shelving, beside a couple of fresh carcasses hooked to the ceiling.

We drove along sandy tracks to visit the graves where Ian's parents were buried, their headstones a little way off from the main house, under the shade of a spreading leopardwood tree, in a spot Esther picked out before she died.

'Mum picked the side that would get the morning sun.'

The stone was brought from Kara, Aubert Jackson's original property, in a single block and when Ian dropped it on the ground it split into two near-perfect halves. He carved the inscriptions himself and he always stopped to say g'day when he passed.

'I need to go over it again, bring it back up to white,' he said, gently brushing the stone.

We saw the stones and posts that marked the spot where Harry Jackson built his first humpy as a lone 23-year old, then the remnants of the first shearing shed that stood alongside the larger shed Ian's father built, using sleepers from the old railway line that once ran between Broken Hill and Silverton.

'He cut all those sleepers himself with a handsaw then drove them three feet into the ground.'

The timber posts came from Broken Hill's north mine and the magnificent Oregon trusses were recycled from an old blacksmith's shop. They held up a roof made of iron recycled from a turkey farm in Broken Hill.

'Dad had to get it at night, when the turkeys were asleep'.

The old shearing shed was an evocative place that had gradually been left to decay. While a hot wind raged outside, it was near-silent inside, as if time had been suspended. Dust motes hung in photographic stillness and shafts of sunlight poured through broken timbers. It was easy to imagine the noise and fury of shearers hard at work in the hot, cramped conditions of decades past.

The dusty, uneven floorboards; the puncture marks in sheets of corrugated iron; a wasp's nest in one corner and a willie wagtail nesting in another – it all combined to give the impression of a movie set, an old John Wayne movie set no less, cowboys and Indians slugging it out for supremacy.

I wasn't far wrong. In the early 1980s there was a bitterly contested six-week shearers, strike across Australia, aimed at preventing the use of new combs. Farmers wanted to introduce wider combs to increase productivity but shearers feared the new equipment would open the floodgates to cheap labour from New Zealand, where the new combs where widely used. At the height of the wide-comb dispute a rifle was kept in every shearing pen at Tirlta, vital protection against those opposed to the new combs. As soon as shearing finished, Ian's father put the shearers in a plane and flew them to Broken Hill.

'He didn't even tell air traffic control.'

The enormous old wooden plate (often called a monkey) that was used to press wool packs was rusty but it still worked, after a fashion. It was no wonder Ian and Merry's daughter, Sarah, used the shearing shed as the backdrop for her wedding photos when she got married on the property in 2005.

It was like walking through a 3D history lesson. Ian threw nothing away – a policy that I suspected had as much to do with the importance of family history as with any conviction that he might one day find a use for the object in question. There was a story attached to everything, from the clipping shears his grandfather had owned to the wire fencing his father had been forced to re-use due to shortages after World War II.

He showed me the 340-foot-deep well that his father had dug near the house, using a pick, shovel and crowbar, interlacing the sides with timber as he went down to stop sand coming in. It was all steeped in the sweat, effort and achievement of three generations of the Jackson family.

In keeping with family tradition, Ian built his own shearing shed, working with his brother to peg out the corners of a floor area the size of an aircraft hangar. It was an enormous structure and I wondered if he needed planning permission for a place that size. Silly question. Who needs planning permission when you don't have a plan?

Ian designed the 50-by-156-foot-long woolshed to accommodate cyclone-rated trusses from the Shell Oil Company in Broken Hill, which he'd had the foresight to buy when the building closed down in the 1970s. Almost twenty years later, he used the trusses to form the basis of his new shed.

The foundations were made from cut-down 44-gallon oil drums that were sunk into the ground and filled with concrete. Ian waved a stick around inside to make sure there was enough height when handling sheep.

When Ian switched on a shearing comb to demonstrate its use, the noise was deafening, although not as loud as a sudden unexpected thrumming that startled me.

'What's that?' I asked, alarmed by what sounded like a plane buzzing the roof.

'Just a willy-willy.'

Dust devils were a frequent phenomenon in the bush, touching down suddenly in a vortex of dust, but the mini-tornado that came through just after Ian had erected the frame must have been far more powerful than any dust devil.

Merry had been out riding with their daughter soon after the mini-tornado blew through. Sarah pointed to the newly erected frame.

'Mum,' she said, 'why did Dad build the woolshed so close to the ground?'

The tornado had knocked the whole lot down, forcing Ian to start again, and he worked through the burning mid-summer heat to complete the building.

'The whole thing cost me just $60,000,' he said with pride.

Merry cleared away the candlewick quilt she'd been working on at the dining table to serve a lunch of quiche followed by traditional flummery (two lemon jellies, water, egg, sugar and fruit; it was seriously delicious).

'I didn't fall in love with Ian, my father did.'

At the time, and for several weeks after my visit, I thought Ian's wife was called Mary; it was only later that I realised how well the name Merry suited her.

'I wasn't very nice to him at all, isn't that awful? But he's gorgeous, he really is. They grow on you.' Her clear voice rang with kindness and she laughed with gentle grace as she told me the story of how they met.

And no offence, Ian, but I'm not surprised she wasn't very nice to you.

*

It was a Saturday evening in late October 1972 and Ian was at his desk at Marcus Oldham College in Melbourne, surrounded by books. Final exams were looming and he was determined to do well. Since the trip to America, Ian had knuckled down and adopted a disciplined approach to study. His reward would come later, on an overseas trip to work on farms in Canada, France and the USA. He'd already booked the ticket and planned his itinerary.

When he got back he would have the option of working for Sandy Reid again. Sandy had employed Ian jackarooing on Narrangullen near Yass during his mid-year in college and Ian had responded well to the man's tough work ethic. Narrangullen was a 'big joint' with 15,000 acres, 1500 sheep, 1000 cows and 2000 or 3000 acres of cropping. Then again, set against working for Sandy was the lure of being his own man. It was a choice Ian would have to make when he got back from overseas, for now all that mattered was getting through the exams.

'Ian?'

He recognised his mate's excited voice on the phone. 'Hey Piggy, what's up?'

'I won at the races. We gotta celebrate!'

'Sorry, I've got an exam tomorrow.'

'Mate, you don't understand. I put down ten bucks and the horse came in at 110 to one. The drinks are on me!'

For a student in those days $1100 was a huge amount of money; all Ian earned was a measly ten dollars a day. He felt torn. He wanted to join his mate but he couldn't afford a big night at the pub. In the end, he compromised.

'I'll join you later. Just before the pub shuts.'

In another part of town, a quietly spoken nursing student was under pressure to join the party for different reasons.

'Merry, please, *please* let's go, it will be so much fun.'

The thought of loud music and a drunken crowd of agricultural students held zero appeal for eighteen-year-old Merry, who led a quiet life and was considered the calmest one in the family, but her older sister Lynor was longing to join the party. Lynor knew all the Marcus Oldham boys and she loved parties. The problem was she had been grounded and the only way her parents would let her go was if Merry, her sensible younger sister, went with her.

'Pleeeease, Merry?'

Merry knew if she didn't give in she'd be forced to spend the rest of the night listening to her sister pleading and complaining. She sighed and went upstairs to get ready.

Ian walked into the pub at quarter to ten, fifteen minutes before closing time, and sat down with Piggy and his mates just as Merry and Lynor walked in. The two girls were introduced to everyone at the table and Ian turned to his friend, Eric, who was sitting beside him.

'See that girl Merry? I'm going to marry her.'

'Hah! You'd better get her to talk to you first!'

Merry had no idea of the effect she'd had on Ian. She congratulated Piggy, spoke to a few of her tennis friends and obliged her sister by staying until the pub closed but by then she was ready to leave. She hated noisy crowds.

'Come on, Merry, we're going dancing!'

Lynor grabbed her sister's hand and before Merry could object she was dragged next door into the dance hall.

Ian paid the two-dollar entry fee and followed them. He bought Merry a drink and asked her for a dance, but the eager lad with a slight limp made no impression on Merry at first. It was only later, when it was time to leave and Lynor was nowhere to be seen – as often happened – that Merry paid Ian any real attention.

'Have you seen my friends?' she asked, scanning the dance floor for a familiar face.

'They've gone.'

Merry frowned. 'Why would they leave without me?'

Ian grinned (big mistake). Earlier that evening he had made it his business to find out who Merry's friends were. Then he'd approached them, one by one, and told them Merry wouldn't be needing a lift home that night.

'She asked me to let you know,' he said. As a result, they had all gone home.

Ian stood in front of Merry and his grin widened. 'I told them I'd be taking you home.'

Merry was furious. 'How dare you!' she said, itching to slap his smiling face.

Ian wasn't concerned that he'd offended Merry. He knew that with no other way of getting home she would be forced to accept his offer of a lift, and that meant he would achieve his objective of finding out where she lived.

The next day he called into the travel agency and cancelled his overseas ticket.

'Why?' the bemused agent asked. 'You only booked it two weeks ago.'

'Getting married.'

'Hey, congratulations!'

Once he had her address it was easy for Ian to obtain Merry's phone number and over the next few days he rang the house several times a day. Each time Merry refused to speak to him.

'I'm sorry, Mr Jackson, I'm afraid my daughter is unable to come to the phone right now, would you care to leave a message?'

'Please tell her I rang and I'll call again.'

When Merry's resolve showed no sign of weakening, Ian drove to her house and parked outside. Merry's father, Stan, was in the front garden, perched on top of a ladder propped against a tree as he struggled to lop off a large branch. It was too good an opportunity to miss for someone like Ian. He got out of his car, rolled up his sleeves and sauntered across. 'You look like a man who could do with a hand,' he called out.

Inside the house Merry and her mother were preparing lunch. Betty glanced out of the kitchen window. 'Who's that lad helping your father?' she asked.

Merry was horrified when she realised who it was. 'Mum, whatever you do, do *not* let that man in the house!'

Out in the garden the two men chatted while they worked and Stan was delighted to discover Ian had been raised on a sheep station out of Broken Hill. His own stint as a roustabout near Menindee at the age of sixteen had left him with an abiding love of the bush, in particular that part of remote New South Wales around Broken Hill.

(All I can say is, Ian Jackson, you were one lucky bugger.)

In the end Ian came clean and confessed his interest in Merry.

'I think I got off on the wrong foot at the dance,' he said.

'Come in and have a drink, lad, I'll have a word with her,' said Stan.

Faced with her father's insistence that Ian was a decent lad, and that he was sorry for the way he'd behaved, Merry had no choice but to be polite to him.

The next time Ian rang, Merry's father held out the phone. 'Give the lad a chance,' he said. 'He's got a good head on his shoulders. At least go and have dinner with him.'

Less than twelve months later they were married and nineteen-year-old Merry moved from the heart of Melbourne to Tirlta, a windswept sheep and cattle station in Outback New South Wales.

I got the impression they both enjoyed recounting the story of their unlikely start.

'Ian was a nice man and he behaved well on our first date but he doesn't now,' she laughed. 'Ian is cheeky and it's good to have fun but when we're working hard and things are serious we buckle down and get on with it.'

The 1970s weren't an easy time to be a grazier. In their early years on Tirlta, Ian and Merry lived in one of the old shearers' quarters, receiving beef and lamb from the property and earning an additional twenty dollars a week from shooting foxes. Their first child, Matt, was born in 1976 and when Merry had difficulty breastfeeding Ian approached his mother for help to buy powdered milk.

'She refused, said I had to make my own way. That was her way of showing me how tough life was.'

Ian went cap in hand to a friend in Broken Hill to ask for a loan, a sobering reminder of their precarious financial position, but Ian had no doubt they would survive. Why, I wondered? He credited a film he saw as a child, the title long since forgotten.

'It was an old black-and-white war film and I watched it in my first year at high school.' One scene left a lasting impression. 'A woman's husband was taken out and shot in front of her. She had a little child and she did everything she could to try to survive. In the end the bailiffs took her house away.'

Ian vowed that would never happen to any family of his.

The seminal year in America had laid the foundations for a phenomenal work ethic, and a near-death experience following a car crash in Ian's early twenties – with the classic white light and feeling of peace – reinforced his determination to work hard *and* play hard. The experience also helped explain why Ian admired the larrikin in others, and why he nurtured it so much in himself.

Within six years of moving to Tirlta, he and Merry had three children – Matt, Andrew and Sarah – and they worked as a team, often taking all three children with them when they mustered on motorbike, the youngest tucked into the front of an overcoat.

There was a lot of 'buckling down' in those days. Falling wool and stock prices forced Ian's brother Murray to take work in the Broken Hill mines and Ian helped run his property while he was away. Ian and Merry lived hand-to-mouth, paring back costs wherever they could as the situation worsened. It reached the point where Murray was forced to sell and Ian agreed to buy his brother's sheep for twenty cents a head.

'It was a terrible time.'

By 1978 it got too much for Ian's father too and, at sixty-seven, Harry Jackson decided to retire.

Ian and Merry scraped together every cent they could borrow in order to buy Tirlta, helped by Ian's payout from the car accident seven years earlier. As sheep prices recovered they bought an adjoining property, Rowena. Interest rates then climbed from ten per cent to eighteen per cent, and with three children at boarding school they were borrowing money at twenty-one per cent to service their other loans. The combined interest rates were almost forty per cent.

They had taken on a crippling level of debt but nothing would have made Ian Jackson buckle.

'We just had to dig in and hang on.'

They got through the difficult years and, as each child finished school, Ian and Merry left them in charge of the property.

'We went away for a couple of months and left them to it, that way they found out what was involved,' said Ian.

Their oldest son was never in any doubt that he wanted to be a grazier like his dad. Matt would wake the governess early so he could get his schoolwork over and get out onto the property. After studying agriculture at uni, he and his wife, Sarah, bought the adjoining 72,000 acres at Rowena from his parents, which is where they now lived and worked with their two young children, Sam and Archie.

Ian and Merry's second son, Andrew, survived an extreme bout of food poisoning during his time in charge, caused by a burger he'd left in the sun on the dashboard of his ute. While his parents were away, Andrew successfully negotiated the sale of several thousand sheep for more than they were worth, achieving $50 a head when their more experienced neighbour only managed $49.51. His eventual decision to work as a wool buyer in Melbourne can't have come as a great surprise to anyone. He set up his own business with colleagues in the town his great-grandfather came from, and where he now lived with his wife and two children.

Ian and Merry's third child, Sarah, began her career as a para-medic in Broken Hill, always hoping she too might one day work on the land. It was a dream she shared with her husband, Luke, who at that stage worked in the mines. When they found a suitable property to lease, Australia was in the grip of a prolonged drought – not an ideal time to give up well-paid work – but some instinct told them to take the plunge. Less than a fortnight later, the ten-year drought broke. When they eventually bought Katalpa, a 140,000-acre property north of Broken Hill, they were convinced they'd have to spend tens of thousands of dollars on a water pipeline. They sat down to do the sums one night and woke up the next day to find three inches of rain had filled their empty tanks. In two days they received the equivalent of eighteen months' worth.

'Sarah seems blessed with good luck,' said Merry.

'Kissed on the arse by a fairy that one,' said Ian, characteristically more blunt.

Ian, a 62-year-old grandfather of seven, was justifiably proud of his family's achievements and showed no sign of slowing down. When (and if) he ultimately did, he was planning to do some writing.

'I'll write one book about my life and another about the bastards I've met along the way!' he declared.

Ian and Merry had reached an age when they could start to take things easy but neither was interested in moving off the property. When Sarah and Luke were looking to buy Katalpa Station they asked if their parents might be interested in retiring.

'But we love what we're doing,' said Merry. 'And where would we go? It feels lonely just visiting a city.'

Others echoed my impression of Merry as a quietly selfless woman, universally loved and admired by all who met her.

'If anyone deserves an award for inspiring people, my mother would be the number one candidate,' Andrew said at her sixtieth birthday celebrations in 2013.

When the last bad drought hit, Merry and her friend Viv Turner organised a regular get-together for women in the area, ostensibly to learn quilting and sewing, in reality to talk through issues that affected all of their families.

'The biggest single factor during drought is isolation. We have some of the best ground to operate on, so if we're struggling, we know that everyone is struggling. Our nearest neighbour is sixteen kilometres away and we know more about what they are doing than someone in a city who lives six metres away from their neighbour.'

Out in the bush, they were all in it together. They suffered the same dust storms, the same droughts and similar freak weather events. The flood that inundated Merry and Ian's property in February 2000 though was exceptional.

*

It was a warm night and partygoers at George Siemer's twenty-first birthday party drifted out of the woolshed, heading for an elevated loading ramp. Ian and Merry were among the group gathered at Koonawarra Station, two hours north of Broken Hill, as were Lynne and John Gall from neighbouring Langawirra. The Galls joined others in commenting on a thin band of high cirrus cloud drifting in from the west, partially obscuring the moon. The party rolled on; rain, if it came, was always welcome.

Most people were planning to swag it but Ian and Merry needed to get home. They had guests staying and wanted to see them off the next day. The Jacksons drove forty kilometres south to Tirlta and the Galls drove twenty-five kilometres north to spend the night at Nuntherungie with John Gall's brother.

Ian and Merry woke next morning to a darkening sky. An inch of rain quickly fell. The rain clogged the sandy track and their departing friends got bogged, forcing them to turn back. By early evening it was clear a big storm was brewing.

Over at Nuntherungie conditions deteriorated too and the Galls decided to skip lunch and head home. Rain fell steadily as they pushed through slippery mud, sliding on dirt roads that were soon underwater in places. Progress was slow and when they reached the fast-running Coonigan's Creek they had to sit in the car and wait, hoping levels would subside.

Sticks and stones laid at the high-water mark quickly disappeared as the water crept higher and it was soon obvious that they would never make it across. They drove back to Koonawarra only to find the homestead cut off by Nuntherungie Creek. Nick Siemer boated across to pick them up and they spent the night on a mattress on the dining-room floor, made up by Nick's wife Ruth and his daughter Fleur.

Back at Tirlta rain fell in thundering torrents all night. When it showed no sign of stopping, the Jacksons grew anxious. Tirlta sat right in the middle of the Bancannia trough, between two sets of hills that lined a massive gully. Water ran off both sides, which made for good rich soil but left them vulnerable to flooding.

Early on Monday morning, Ian radioed Koonawarra. He knew John and Lynne Gall's property would be affected and he wanted to warn them.

'Is that Lynne? How much rain did you get over there?'

'Three and a half inches, isn't it fantastic?'

'You'd better sit down. We've had ten.'

'You're kidding.'

'No, I'm not. Put John on.'

Ian heard Lynne call her husband to the radio.

'What's all this about rain?'

'Mate, you need to get home. There's a heap of water here and it's heading your way.'

The storm that broke that day in February 2000 was a freak event, so localised that the Jacksons' nearest neighbour only received three and a half inches. Friends on the other side of Broken Hill didn't even get half an inch. On areas around Tirlta and Langawirra an astonishing eleven inches fell in six and a half hours; at one point the floodwaters were twenty kilometres wide and a hundred kilometres long, submerging low-lying paddocks under ten feet of water.

It was hard to believe that the dust bowl I had driven through earlier had once been submerged under ten feet of water. It sounded incredible and of course turned out to be true (I checked).

When it came to weather in remote areas it was frequently a story of feast or famine, people's lives governed by the unstoppable forces of fire and flood or dust storms and drought. I sometimes wondered if that explained why bush folk seemed less uptight than city folk, more willing to accept that the forces controlling their lives were just that – in control. Or maybe that was another romantic folly of mine?

The station homestead at Tirlta had grown organically over the years, starting with a lone tank sinker's hut. Tank sinkers worked during the early days of settlement, travelling from station to station with bullocks and camels, spending months at a time carving out water tanks on stations without running water. (When Ian told me they had several water tanks I assumed he meant the kind of water tank we had in the back garden. His tanks were vast reservoirs thirty feet deep.)

When tank sinkers moved on they usually left their modest huts behind, and the same was true of boundary riders. Ian's father had built up quite a collection of huts and they formed the basis of the house. When he got back from fighting in Borneo he continued to expand the collection, so, by the time Ian and

Merry moved in, the homestead consisted of several discarded huts joined together on old mulga posts.

It also incorporated the remains of the disused Mutawintji Hotel (a one-time Cobb & Co staging post in the early 1900s). Harry had bought it then dismantled it piece by piece, pulling out all the nails from the timber and marking the irons for the roof so he could re-use the hotel as part of his ever-expanding Tirlta homestead. Recycling writ large.

Incredibly the homestead at Tirlta was spared from the flood, but the impact of eleven inches of rain was too much for the quirky building and it sagged and sank beyond repair.

Rebuilding the house gave Ian and Merry an opportunity to plan a new layout which included, for the first time ever, an inside toilet. 'Merry was tired of having to fight off spiders in the dunny,' Ian explained.

They ripped the side off what remained of the old house the day before a massive dust storm hit. Plastic sheets tacked onto the side of the building kept some of the dust out but friends who'd arrived to visit were horrified.

Ian responded to their concern with his usual self-confidence.

'If it was good enough for my grandparents, and my parents, it's good enough for us.'

Merry dug in with quiet tenacity.

'We'll get there,' she said.

Over the next eighteen months they suffered a parade of near-continual dust storms that slowed progress but never halted it. Merry worked as hard as Ian did and they would sometimes shear from six o'clock in the morning then head back to work on the house at night, frequently not going to bed until one o'clock, rising a few hours later to start work again.

But they got there, as Merry knew they would.

Their new house was impressive, a gracious L-shaped building facing northwest with steep overhangs to shield occupants from the scorching summer sun. It was finished in time for Sarah to hold her wedding in the garden in 2005, a garden that was partly created, like so much on Tirlta, from recycled materials, including posts reclaimed from the horse yards that Ian's father had built. They defined the boundary and formed the basis of new garden beds.

Like his father before him, Ian incorporated the Mutawintji pub into the redesign, using the original four rooms to create new living and dining rooms. He pointed to an unfinished spot on the pressed metal ceiling.

'I left that as a mark of respect to my father.'

And the spacious new bedroom wing had not one but two indoor toilets.

'I thought, ah bugger it, I'll build another bedroom and an ensuite.'

The house was undeniably beautiful, with floorboards of spotted gum downstairs and Western Australian sheoak upstairs. A cross breeze cooled the multi-purpose eyrie that served as a lookout for fire and flood, a quiet place to savour a drink at the end of a long day and enjoy the sunset, watching colour shift on the distant Barrier Ranges beneath the ever-changing night sky.

One thing that didn't change was the kitchen, still resplendent with bright green cabinets from the early 1980s.

'I've always liked this kitchen. It's a welcome burst of colour when you come back from a dusty paddock,' said Merry.

To celebrate completing the house they splashed out on an antique dining table made of Queensland blackwood, originally commissioned as a boardroom table for Harris Scarfe in 1916. It was large enough to seat twelve people with ease, and at their first Christmas in the new house they rejoiced at seeing both families come together, including Merry's parents, their children and associated offspring, and all of Ian's family.

Recalling that family Christmas, Ian embarked on a tale involving the Emperor of Tokyo. Following his train of thought felt a bit like running to catch a bus as it disappeared around a corner.

'The Emperor of Tokyo got sick and Dad's favourite food was crayfish. The price crashed because no one in Japan would eat crayfish in sympathy and Australian crayfish normally went to Japan so prices dropped to eight dollars a kilo and I'd sold a bit of wool so we flew to Kingston and brought back sixty-eight kilos.'

Sixty-eight *kilos* of crayfish? Ian had to retrieve several wayward fish from behind the pedals and under the seats of his light aircraft on the 1500-kilometre flight back. He dropped batches with various mates and delivered the remainder to Merry, who cooked them, rinsed them in freshwater and hung them by their tails on the clothesline.

'They were bloody beautiful. The old man was in heaven. That was the most beautiful Christmas you could poke a stick at.'

It must have been disheartening for Ian and Merry to have to rebuild their deluged home, but water is the lifeblood of the bush. The floodwaters that ruined their house also filled the ephemeral lake on Tirlta, so on New Year's Day 2001 – when the temperature hit fifty-two degrees in the shade – they spent the day at the lake.

'Let's take Deb out there,' said Merry. 'Would you like to see it?'

'I'd love to.'

And that was how I discovered Ian Jackson's secret.

*

The sandy track we followed cut through fertile land peppered with acacia trees (including nelia and hard, twisted mulga), black oak, bully bush and, according to Ian, 'eighty different varieties of saltbush'.

Over millions of years the gully on which Tirlta stood had filled up with rich dry silt, trapping vital water supplies 300 feet underground. It was an area steeped in history, and not just the history of Ian's family who first settled in the area in 1888.

The rocky gorges and sandy creeks of nearby Mutawintji were rich in sites of cultural significance for Aboriginal people. Legend had it that 'Tirlta' was a jealous old man who killed his intended bride when she spurned him, along with every other member of her tribe. When Tirlta returned the following season he searched in vain for the bones he knew should have been there. He could find, according to CP Mountford's 1965 book *The Dreamtime*, only 'carpets of scarlet flowers with black eyes that had grown from the blood of the slain'. Those scarlet flowers were the elusive Sturt's desert peas or Flowers of Blood, and they grew in abundance in the area, especially on the adjoining Langawirra Station.

'This country is deceiving, when it's good it looks fantastic but after ten years of drought you see the true colours of the bush.'

Ian stopped the car and he kicked around in the dirt for a while, unearthing the stone-like remnants of a white-ant nest that had been used on a fire.

'They hold the heat and glow like briquettes on a barbecue. You find nardoo stones around here too. Aborigines would collect nardoo seeds and crush them to make damper. The stones they used to crush them were too heavy to carry around so they'd leave them in certain spots.'

As someone who used to be on the board of Mutawintji National Park, Ian undoubtedly knew what he was talking about.

Fifteen minutes later we pulled up on the shores of a lake so wide, so blue and so beautiful it literally made me gasp. The clear stretch of azure water superimposed across the red dirt land-scape could have been an ad for the Australian Tourist Board.

'We're lucky. How many people can go home after work and enjoy this sort of peace?'

Not many.

'Beer?'

Ian handed me a stubbie.

'If it were permanent I'd have a house here,' said Merry. 'Out of thirty years we've only had twelve with water.'

What a treat to have seen it. In the short time I stood watching, two white swans swam past, several pelicans flew overhead and a group of ducks bobbed in the weeds by the shoreline. Black dots on the opposite bank could have been goats or feral pigs; it was impossible to tell across such a vast distance without binoculars.

When full the ephemeral lake at Tirlta was a favourite spot for the family to meet for a drink before sundown. Ian and Merry's son Matt was there that night, with his wife Sarah and their two sons, eight-year-old Archie and ten-year-old Sam, plus James, an English backpacker who'd been working on Rowena.

The boys clambered out of the water to introduce themselves and their father wrestled them to the ground, tickling them and rubbing dirt in their hair until they were helpless with laughter.

'You got a bullshit filter on that tape?' Matt asked.

Ian answered before I could. 'Hah! That would wipe out the lot.'

Father and son clearly got on well.

Young Sam was a confident lad, convinced of his potential to play footy for Geelong, which was what he planned to do before settling down to work as a grazier.

'Don't be too disappointed if you don't get in, will you?' said his grandfather.

'Nah, even if I only get to VFL it will be good.'

'There's always something else you can do.'

'Like cricket,' said Sam, his face lighting up.

The two young boys were full of fun and, like most bush kids, they were happy chatting to an adult. With memories of Ian's governess fresh in my mind I asked them about School of the Air.

'It's good, we've got Karen from Scotland next term and we just had a Pommy governess.'

'Some teachers you like more and some really push you,' said Archie.

'We had one arrived at five at night and she left at seven next morning. Lived in Byron Bay. She was crying when she left, dripping with tears. Her word was it was "too rural" for her.'

I had to laugh at Sam's turn of phrase and his expert timing. He'd inherited his grandfather's gift for storytelling.

Ian turned to me. 'You asked me earlier what my legacy was? These kids, that's my legacy.'

Family was everything to Ian and Merry and that family extended far beyond their own three children and seven grandchildren. As we relaxed by the shore of the lake that night there was talk of a trip to America so Matt, Sarah, Sam and Archie could meet Ian's 'other' family – the one that had hosted him in America all those years ago.

Then there were the hundreds of children from around the world who had stayed at Tirlta as part of the American Field Service program. For twenty years Ian and Merry offered an outback experience that proved as transformative for some of the foreign students as Ian's trip to America had been for him.

'It was the equivalent of sending a bush kid to Times Square, New York.'

Once they overcame their initial nerves and the horror of blowflies buzzing over long drop toilets, city kids from America, Spain, Japan, Canada, Chile, Bolivia, Finland, Thailand and beyond were first in line to muster calves or test cows for pregnancy, arms buried up to their shoulder blades inside a cow's womb.

The most emotional moments occurred on the shores of that ephemeral lake. Ian would drive the students out at dusk, the sky darkening above them as birds called across the water, and drop them at fifty-metre intervals along the shore. Each student would be left alone for half an hour, with a torch for emergencies and an empty can of soft drink to mark the spot. The rule was no talking and no smoking.

Dumbfounded by the quiet and the bush noises they began to hear once their ears had attuned to the silence, the students saw satellites, shooting stars, falling stars and the brilliance of the Milky Way.

'That's how they learnt to appreciate the true nature of the Australian bush. Some of them were in tears when they left.' Ian's voice was thick with emotion at the memory of such transformative experiences.

'Merry, are you prepared to do the lift?' he said.

'Okay,' said Merry, quietly. The sun had set by now over the aquamarine lake and it was getting hard to see people's faces.

'James?'

'Okay.'

'We're going to do it to you,' said Ian, approaching in a way that suggested it wasn't worth protesting.

He sat me on a chair and positioned people at each corner: James and Matt in front, he and Merry behind.

'We're going to lift you using just two fingers. We'll have a trial run first.'

There were jokes about how much I weighed and the chair wobbled as they tried to lift it with their hands. Ian silenced them.

'We're going to build a stack above your head. Concentrate, everyone, this is serious.'

And suddenly it was. Conversation and laughter stopped as hands moved silently above my head.

'She's as light as a feather. Concentrate now.'

The air felt charged with something, a sense of purpose perhaps (or my nervousness). The stack built slowly, it was just as slowly removed then fingers were tucked under my armpits and . . . *Whoosh!* I was propelled several feet up into the air, as easily as if gravity no longer existed.

'You were so light,' said Merry, laughing with glee when I was back on solid ground.

How did they do it? What had happened to all the weight they'd struggled to lift earlier?

'You asked me how I lived my life. You've just seen it,' said Ian, eyes sparkling in the darkness. 'Energy, focus and concentration; you can do anything if you believe you can.'

*

After such a mystical experience at the lake, I woke up next morning worried that Ian's collection of memorabilia might prove disappointing. I should have known better.

Most of the valuable items of historical significance were kept safely under lock and key, but tucked away on a part of the property few people visited was an old tin shed crammed full of fascinating pieces. We squeezed past a fence, cracked open the locked door and stepped into a dusty treasure trove of such magnitude it would have made Jack Sparrow's eyes light up.

Thick layers of dust had taken the gloss off some of the shinier pieces of glittering stone, and there was nothing of much financial value, but no amount of dust could hide the beauty and wonder of that remarkable collection of Australiana. From simple glass bottles to intricate medical chests, from bottled lizards and box cameras to smoothing irons and shards of potch, Ian's shed contained a wealth of history.

Ian had always been taught that if he wanted to know something he would have to remember it. His father never wrote anything down. Instead he would walk his sons through the makeshift museum, telling them stories that had been handed down from his father, who began the collection back in the late 1800s.

Ian wove a spell as captivating as the many artefacts that lined the walls and filled the glass cases.

'That flag was carried in Borneo,' he said, pointing to a tattered scrap of Australian flag standing limp in one corner.

'See that water bottle? It was carried in the charge of the light horse brigade in the Battle of Beersheba.'

The water bottle stood next to an unopened ration pack from World War II, which was next to a Morse code mirror kit in perfect working order.

There were fascinating clues to Outback life in ages past and no reason to doubt the veracity of Ian's stories once the shock of finding such treasures locked away on a dusty sheep station had faded.

He pointed to a fan with a two-stroke piston engine that ran on methylated spirits.

'When my grandmother Ida fell sick in 1919 my grandfather bought her that fan.'

The fan would have blown air onto a damp sheet hung across a doorway, helping the sick woman cope with the fierce heat.

There were shelves of smoothing irons that would have been filled with hot coals, others that ran on petrol and yet more with removable heads, piled up beside butter churns and carbide lamps from a coal conversion car.

Lifting the lid on a plain wooden box revealed it to be a 'newly invented improved magneto-electric machine for nervous disorders', with instructions to 'place the metallic wires on any part of the patient through which it is desired to pass the electric current'. Turning the crank handle at varying speeds regulated the strength of the current. 'It is less unpleasant to the patient if wet sponges are placed in the ends of the handles.'

Bottled spiders were lined up beside king brown snakes. There was a stuffed alligator, an empty turtle shell and a box camera lost in the floods of 1956 then retrieved last year when high winds uncovered it from beneath a sandhill. A Trafalgar cold safe that would have kept food cool by trickling water down its canvas sides must have been a prized possession. It was still in pristine condition.

Ian was quick to point out that all the Aboriginal artefacts were collected from the 1880s to the 1970s, knowing the practice was now illegal. There were shoes made of emu feathers and human hair worn by a kadaitcha man; killing boomerangs and spear tips; emu claw necklaces; ceremonial belts made from human hair; nardoo stones and polished flints; animal shells; carved boab seeds; fertility dolls and exquisitely decorated emu eggs.

'I'm preserving all this for my grandkids,' Ian said, brushing red dust off a glass case. 'I bring them in here each time they visit and I tell them the stories.'

Subtle changes must have crept into the telling and retelling of those stories and I wondered how much more they might have changed by the time 'the grandies' as Ian called his grandchildren were old enough to tell their children.

They were in drought again on Tirlta, like so many in that part of outback New South Wales. In the last big drought, Ian had planned to send his cattle to Queensland. He was walking them into the yards in preparation when a massive dust storm hit, and within minutes the sky had turned from dark orange to purple then black.

'I couldn't see the person sitting next to me.'

Power lines went down, phone lines were knocked out and even the VHF radio failed as the storm raged for thirteen hours.

The following day Ian and Merry took stock of the damage. Leaves had been stripped from trees and bushes, cattle had been felled where they stood and sheep that sought refuge in dams were bogged and buried alive.

'We were lucky though. Some people west of us lost thousands of sheep and cattle.'

Drought is a terrible thing for any farmer or grazier.

'You hope things will come good and you have to make decisions that hurt at the time but it's better to see a little bit of credit on your statement than none at all and bones in your paddock. I've been guilty of holding onto cattle for too long, then shit you're buggered.'

The Jacksons knew they were fortunate to live in a fertile valley with plenty of scrub and saltbush that protected them from the worst of any dust storm. They also had a permanent source of water, even if it was hundreds of feet below the surface.

The biggest clip of wool they'd ever had at Tirlta was 860 bales. During the last drought that dropped to fifty.

'The saddest story I heard was from Ivanhoe. Shearers would normally spend three weeks there. In the last drought they finished the cut-out by nine-thirty on the first day.'

Before I left Ian and Merry showed me a framed poem, written by their daughter Sarah and circled by handprints from their seven grandchildren.

Take a look around you and what do you see?
A station, a house and a growing family,
On these labours of love so much sweat has been shed,
Many hopes almost lost, many dreams have been fed,
Ian and Merry have travelled so far,

Deb Hunt

From awkward beginnings in a Victorian bar,
So relax in this abode that together they built
And let their warmth surround you like a feather down quilt.

I couldn't have said it any better.

Cath Marriott

'Yarallah'
Ten kilometres west of Benalla, northeast Victoria

Searching through the archives of the Rural Woman of the Year Award I came across the details of a young woman in Western Australia whose passion for the land seemed evident in everything she did.

I rang Catherine Marriott to explain the premise for this book and she said, 'I'm not the one you should interview. It's Mum you should talk to.' Catherine gave me a few details of her mum's story and I had to agree. 'The only problem will be convincing her,' said Catherine. 'If she's reluctant, tell her I said she should do it.'

Sure enough, her mother, Cath, hesitated. 'Why me?' she asked when I rang. 'I haven't done anything special.'

By that stage I knew that wasn't true. Cath raised four children alone after her husband, John, died of stomach cancer at the age of forty. She kept their farm going and survived the catastrophic effects of drought, debt, fire and flood – the latter two events leading at one time to the destruction of their family home and the loss of most of the flock. In all that time, Cath never once thought of giving up and selling.

'Catherine said you had to,' I said, trusting her daughter's instinct.

'Oh, did she?'

We spoke on the phone for a while, briefly discussing the need for greater awareness of food production and the importance of organic farming practices, and Cath's reserve gradually gave way to enthusiasm.

I discovered later that reticence and lack of self-confidence were intrinsic to Cath's nature, coupled with burning curiosity about the world and a spiritual connection to land and all living things. Our short conversation was enough to convince me that Cath would definitely be worth interviewing and I was delighted when she agreed to participate.

'You must come and stay,' she said, with the same humbling generosity I had encountered everywhere during the writing of this book. 'The house isn't finished yet so there's no door on the bathroom or the toilet but you won't mind that will you?'

I arrived at Yarallah early one winter's evening on a day of tremendous rain, exhausted after two days of interviews followed by a five-hour drive. All I wanted to do was drop my bags and fall asleep.

Then Cath started talking and I was captivated, all thought of slinking off to bed forgotten. We talked for hours, on subjects as diverse as the world of spirit and child rearing, sustainable food production, biological diversity. Next morning Cath picked up the conversation exactly where we'd left it, as if the intervening eight hours sleep (although I suspect Cath survived on far less) had lasted no longer than the time it took to put the kettle on.

Sitting on her veranda in the winter sunshine that day, surrounded by mature gardens overlooking paddocks rich with grass that shone from the previous day's rain, Cath told me stories of how she and the children had coped in the years following her husband's death. She described the successful lamb finishing operation they had established, the land they'd regenerated, the trees they'd planted and the thriving holiday rental business they'd recently started that introduced weary Melburnians to a slice of rural paradise.

'This farm is successful only because of the energy and commitment of the many people who've been involved in it,' she said.

One of those people was undoubtedly Cath Marriott.

The gift of life

Cath greeted me like a long-lost friend.

'Let's have a cup of tea, I haven't had one all day. It's been one of those days. Tom's going to Western Australia to surprise Charlie so I went to wood-turning to finish a pestle and mortar for a friend over there. I've got to get it done before he goes.'

I sensed this was going to be another discursive chat about anything and everything, and only later would I make sense of who or what Cath had been talking about. No matter, I was used to it by now and Cath was an easy woman to listen to.

We settled into comfortable armchairs. Newly washed sheets hung in front of an open fire burning a big chunk of boxwood, and heat radiated through the open-plan hub of Cath's house. I listened with rapt attention as she talked freely and openly about her extraordinary life, made all the more extraordinary by her long-held belief that she had little to contribute.

'I always felt like I was a failure, like I wouldn't amount to much. I doubted my own abilities.'

Surely not? It was obvious from the outset that Cath had plenty to contribute to life and I found myself nodding agreement as she talked about avoiding antibiotics, eschewing chemicals, embracing recycling, improving the carbon content of soil, extending environmental planting and respecting the life of the sheep they bred.

'I don't have an issue with growing lamb. They have a good life with me and they know Cath's the silly old chook that chats to them. I feel an empathy with them, which doesn't mean I'm not happy to eat them. The end of their life is as quick, painless and stress-free as possible.'

She picked up another heavy chunk of boxwood and threw it on the fire, showering sparks onto the polished concrete floor, and I was struck by her phenomenal strength. It wasn't so much her physical strength (although that was undoubtedly there) as her emotional strength; Cath's energy was open and loving, a female energy in touch with the world of spirit and intimately connected to the land she farmed. She was as much a force of nature as the wind and rain pounding the house.

She cried as often as she laughed during our long conversation and at one point burst into tears when she tried to express the depth of feeling she had for what she called, 'being as one with this extraordinary world we live in. We are careless with it, invariably to make money or gain power, and that's not acceptable.'

I hope I'm not making Cath Marriott sound flaky because she certainly wasn't. She looked to be around sixty and she was a tall, solidly built woman with shoulder-length blonde hair, clipped back from a smiling face. It became apparent later that Cath was an exceptional farmer, an accomplished wood-turner, a fine musician and a gifted artist, and, as well as raising four children and establishing an extremely successful business, she'd gained a master's degree in strategic foresight.

Yet she went through much of life feeling like a failure? I reflected on that later and realised that some of the most accomplished and sensitive women I knew doubted their own abilities, in spite of clear evidence to the contrary. Cath Marriott had more reason than most to be proud of what she'd achieved.

Cath survived the kind of vicissitudes most of us are thankfully spared. She lost her father when she was barely a toddler and her husband, John, when he died of cancer just after his fortieth birthday. Cath raised the children alone – nine-year-old Catherine, eight-year-old Hannah, seven-year-old Charlie and the youngest, Tom, who had just turned five. With their help she kept the family sheep farm in northern Victoria going, surviving flood, fire, drought and debt to emerge from it all stronger and wiser, with a heart so full of love she could sometimes weep when she smiled, both of which she did frequently during my visit. It was like watching rain fall on a sunny day.

It would have been easy for Cath to cling to her children after the death of her husband, and to seek solace in their company, but instead she gave them the gift of an education and she let them go. She equipped them with the skills they needed to spread their wings and fly, telling them they could go anywhere and do anything.

So what did they do? All four chose to study agriculture and they stayed as connected to Cath in adulthood as they had been in childhood.

Her voice thickened as she explained how proud she was of each and every one of them, and there was that rain again, falling out of a clear blue sky.

Cath Marriott was born in June 1948. Her mother, Kay Fairley, worked first as a secretary and then as a cosmetics consultant for Cyclax, one of the first in-store consultants to travel around Australia in the post-war years, offering advice on beauty and make-up in local department stores. News of Kay's arrival would be advertised in the local paper – 'Kay Ekberg will be at Boans in

Perth on the following dates . . .' – and customers were advised to phone for an appointment.

'It would have been a wonderful job for a woman who was keen to broaden her mind through travel,' said Cath.

The job and the travel ceased abruptly when Kay married George Fairley, a farmer from Shepparton in the Goulburn Valley, whose family had been in farming and business in that area for over a hundred years. Within two years they had two children – Cath and Andrew – but Kay's newly ignited passion for farming was to be short-lived. George died less than a week after Andrew's christening, following a heart attack that Cath suspected was caused by the hepatitis he contracted while serving in the Middle East during the war.

George's untimely death forced Kay to relocate her family to Melbourne, 'encouraged' by George's estate, which insisted the farm be sold.

Severed from the farm and the family, Kay travelled extensively. She took the children out of school early when she found special deals that allowed them to cross Australia by train or to visit faraway places like Vancouver, Hawaii, Japan and New Zealand. She and the children once spent sixteen weeks sailing to Europe through the Suez Canal, heading home via the Panama Canal. Kay loved learning and she believed travel broadened the mind. She encouraged Cath and Andrew to become 'citizens of the world'.

It sounded like one long idyllic adventure. Wrong.

'My mother was a fearsome disciplinarian. She believed children responded best to challenge, not to praise. She once said there are people who are good with children, and people who are good for them.'

Kay Fairley was in the latter camp.

'She believed that children needed strict guidelines to follow, and if you put a child down it would pull itself up by wanting to do better.'

Ouch.

Cath's younger brother, Andrew, did precisely that and he went on to become a well-respected solicitor in Melbourne. The sensitive Cath, who loved animals and whose enquiring nature was always trying to make sense of the world around her, was hurt by such criticism. She took refuge in books, only to be admonished by her strict mother.

'Get your head out of that book and pay attention to the world around you.'

I felt myself flinch.

Following three years of study at Burnley Horticultural College, Kay Fairley became a renowned plants woman and a founding member of the Australian Camellia Research Society. Cath and her brother were taken to botanical shows and plant exhibitions, to the herbarium for meetings with eminent speakers and to public and private gardens around the world.

'My mother imported the first mist-propagating system into Australia and had it installed in a glasshouse that she had built at the bottom of the garden.'

Impressive, but how often did she hug you I wanted to ask (and didn't).

In response to her mother's expectation that children should be constructive at all times Cath learnt to play the flute. 'If she could hear me practising she knew I was working.'

Cath was a diligent student who went on to become first flute in the Victorian Junior Symphony Orchestra. She later embarked on a degree in music, not because she was interested in the theory of music but simply because all that practice had made her really quite good at playing the flute.

Cath was keen to point out that, although her mother was

strict, she was also an interesting woman who bequeathed to her children the gift of learning, an understanding of plants and a love of travel. All true, but the gift of loving self-acceptance was one Cath was left to discover for herself.

Cath's upbringing led to a certain eccentricity and freedom of behaviour.

'I had an English friend, who was working as an assistant to Michael Parkinson and she invited me to be a bridesmaid at her wedding at Claridge's.'

Cath flew to London for the glittering event – a rare social occasion for her – and when the friend glibly assured Cath that she could get her a job at the BBC, Cath believed her. She flew back to Melbourne, gave up the work she wasn't enjoying anyway, packed up her belongings and headed back to England. When she arrived in London the promised job had evaporated. She had no money, no job and nowhere to live.

Cath wasn't about to go back to Australia with her tail between her legs so she took a job as a cleaner and found a flat to share near Hyde Park, then she went to Broadcasting House on a daily basis and pestered the BBC. Could she type? Not really. Shorthand? No. In the end they got sick of her daily visits and offered her a job helping to compile a comprehensive A to Z listing of the first radio performances of every classical work ever broadcast on Radio 3. It was meant to be a two-year project.

Cath embarked on the project with relish, immersing herself in the intense research, and when she wasn't working she had a 'rollicking good time' with good friends at Royal Ascot, Badminton and the Burghley Horse Trials.

After two years, the comprehensive A to Z listing of performances had reached Beethoven (which meant they either seriously underestimated the amount of time involved or Cath had seriously enjoyed herself) and Cath's stint in London came to an end. It was time to head back to Australia.

While Cath had been in London her mother had moved from Melbourne to Sydney then back to Victoria, never settling in one place for long. She eventually returned to farming and bought a property at Lang Lang on the shores of Western Port in Victoria. The old title went to the high-water mark and included a mostly non-existent beach that provided valuable habitat along the foreshore.

Cath joined her mother on the farm, content to give up the glamorous 'flim and flam' of London for a more peaceful life in the country where she happily devoted herself to caring for cattle as the years slipped by.

'I never expected to meet someone, and I certainly never expected to marry the boy next door.'

The front door banged open and Cath's youngest son, Tom, arrived in a flurry of wind and rain, his big voice booming across the kitchen as if he was shouting from the other side of a paddock.

'Hey, g'day there!'

Tom now had most of the responsibility for running the farm and he lived in a cottage behind the woolshed. He was smiling broadly as he lugged in another chunk of boxwood for the fire, exuding the same open energy as his mum.

'Anything you need from Benalla?'

'Just some milk please, sweetheart.'

'Beaut. Take care.'

Tom's arrival prompted us to feed the animals (Cleo the dog, Tuppence the cat, at least one peacock, a flock of guinea fowl and a clutch of chooks) before it got dark.

'I locked the rooster up because I'm trying to teach him some manners,' Cath said, stomping through a muddy chook pen. 'Come on, girls!'

The chooks squabbled around her.

'I love my birds and my animals but everything does have to get along. I can't be bothered with things that aren't harmonious.'

Her first year of marriage was anything but.

The 'boy next door' was John Marriott, a quietly spoken deep thinker and a farmer from a long line of highly respected market gardeners in Melbourne. John was a spiritual man who loved animals and he cherished the land. He had a lot in common with Cath.

Cath and her mother had been living at Lang Lang for six or seven years when John knocked on the door one day and introduced himself.

'Hi, I'm John, your neighbour. Have you had any trouble with vandals cutting your fences?'

'No, we haven't.'

Politeness would surely have dictated that there be more to the conversation than that, but if there was Cath can't remember it. What she did remember was a thoughtful man with beautiful eyes.

Cath was twenty-eight at the time and she'd been running the farm for six years. Her brother Andrew had kept trying to introduce her to prospective partners but Cath wasn't keen; she'd always found parties and social chitchat challenging. She loved farming, she loved animals and she was very happy on the farm.

So when John rang one day and made an abrupt announcement, she gave his proposal serious consideration.

'Look, I quite enjoy your company. I've got a girlfriend and she's meant to be coming here on her way back from Europe but if she doesn't arrive I'd like to start taking you out.'

Cath knew the prospect of meeting anyone when she never left the farm was slim . . . and John did have beautiful eyes.

'Okay.'

We both laughed as Cath described their pragmatic courtship.

'I was diving into a pool that I didn't know the depth of but I sensed this man had a wonderful energy.'

It wasn't the first time Cath had mentioned energy. She didn't follow any organised religion as such but she did believe in a higher power, a source of energy that was available to be tapped into when needed, which was just as well. In their first year together it sounded like Cath needed all the help she could get.

'We went to a restaurant in Inverloch to celebrate our anniversary and John told me he nearly left me.'

'Why?'

'He somehow thought I would make the perfect wife but quite frankly I was awful. Growing up without a father I had no idea how to work at a relationship.'

Despite their rocky start, the relationship flourished and within a couple of years Cath and John took over the property at Lang Lang, buying out Cath's mother and her brother. Their first child was born in January 1981 and within four years they'd had four children.

'We got that formula right,' Cath gleefully declared.

Gippsland wasn't the ideal place to develop their shared dream of raising a family that could live off the land – rising values made it difficult to expand – so they sold the Lang Lang property and invested in a larger farm at Benalla, two and a half hours north of Melbourne. They signed the completion papers on 15 March 1985, the day after their fourth child, Tom, was born,

From the start, John was the farmer and Cath the farmer's wife, a fact that surprised me given that Cath had managed her own farm. She gently reminded me that they'd had four children in the space of four years and two months (and I felt suitably chastened).

'We talked constantly about agriculture and John always consulted me before making any major decisions but my prime responsibility was raising the children.'

John was a disciplined farmer and an enthusiastically hands-on father who often took one or more of the children out with him.

'He thought nothing of interrupting a stock inspection to change Charlie's nappy on the boot of the car.'

Sometimes John would pile all four children into the back of their old red Volkswagen beetle and drive them down to one of the paddocks to fix a fence or check on stock.

The property was in a far better state than the house. Mice had taken up residence in the dilapidated building and the kitchen was overrun with giant tiger slugs. The silent invaders entered at night, oozing through doors and windows so badly warped they no longer opened or closed properly. When a four-foot brown snake was found under the broken floorboards in three-year-old Tom's bedroom, Cath and John decided it was time for a new house.

Forced to shelve their long-held dream of building a mud brick house, due to cost, they opted instead for a passive solar design in brick that ticked all the right boxes.

'Building regulations at the time insisted that we spray dieldrin under the slab!'

Cath's tone of voice made it abundantly clear what she thought of that.

Keen to avoid the use of such a toxic chemical (research had shown it could turn up in a mother's breastmilk), John consulted widely and spoke to scientists at the CSIRO in an effort to find an alternative. The local council wouldn't budge. They refused to accept the suggested alternative so John and Cath defied conventional wisdom by choosing not to treat their concrete slab for white ants. In the end, they paid their builder *not* to spray.

I'd never heard of dieldrin and I wondered if Cath and John were being a tad overly concerned. Not a bit. Dieldrin has since been identified as a deadly chemical and its use was phased out in Australia in the 1990s. It's now banned in most countries.

The house went up quickly, fuelled by the energy and enthusiasm of the whole family, and it was only when they'd finished the last coat of paint that John put down his brush and said, 'I feel a bit crook.'

The diagnosis shocked them all: stomach cancer.

John was a slim, fit man who didn't smoke, didn't drink and didn't use chemicals. He grew all his own organic vegetables and he avoided salted meat (research from Japan and Iceland suggested a link between salted meat cured with sodium nitrite and stomach cancer). The only possible causal link Cath could find was the switch from tank water to bore water while the house was being built.

'The bore water had a high level of sodium nitrite, that's the only thing we think might have triggered it.'

With a growing family and a flourishing farm, John had everything to live for. Sadly, after three lots of punishing chemotherapy, the discovery of inoperable seeding in his peritoneal cavity stopped any further medical intervention.

'I would rather die than put any more of that poison into my system,' he said.

An intensive week with Ian Gawler, an authority on mind-brain medicine and meditation, gave John a vision of how to get the most out of the life he had left.

'John did his absolute best to survive to see his children grow up.'

Having dreamt of seeing the children reach adulthood, John lowered his expectations. He fought to make it to his fortieth birthday, sustained by good food, homegrown veggies and daily meditation.

On the days that John struggled, Cath was there for him, reminding him that unless he had bad days he wouldn't know what good days were. It was about choosing to focus on the positive, Cath said. As the disease took hold, Cath inevitably became more involved in the running of the farm.

'There were many things John could no longer do, but he didn't let that loss define him. He still had his family around him and he was the father of that family. He was determined to live with peace and joy, in the full knowledge that he had cancer.'

Defying the medical profession that only gave him six months, John Marriott lived a life of quality for a further fifteen months. He died in March 1990, six months after his fortieth birthday and exactly five years to the day since he and Cath had bought Yarallah. Cath was with him until the end, quietly holding his hand as he passed.

Cath held her breath for a moment before she spoke.

'I came home at five o'clock and Catherine said, "Dad died at one o'clock last night, didn't he?", and I said, "Yes, sweetheart, he did." She said, "I saw a shooting star at one o'clock."'

Catherine was nine years old.

Suddenly Cath was a single parent and a farmer, torn between the needs of the children and the demands of the farm. Much of her energy and ingenuity had to go into parenting, and her own grief at losing the soul mate she'd never expected to find had to be set aside as she focused on raising the children and running the farm.

Drawing on resources she didn't know she had, Cath navigated her way through that dark period.

'If you can overcome difficulties you find a strength within you. Life is challenging, and we all have to acknowledge that. There's no point pretending it isn't. Rising to meet the challenges of life is a gift I believe we can all give our children.'

Maintaining discipline with four young children under the age of ten required considerable ingenuity. Cath was once so exasperated by their behaviour in the local supermarket that she turned the tables on them.

'I parked the trolley in the middle of an aisle, lay down on the floor and threw a full-on tantrum.'

Her drumming heels and loud voice so alarmed the children (and no doubt other shoppers) that they never misbehaved in the supermarket again.

It sounded like Cath's parenting was far more successful than she gave herself credit for, then she told me a story that made me wonder.

Shortly after John's death, Catherine appeared in the kitchen one afternoon with her younger brothers and sisters. They were carrying suitcases that Catherine had packed for them. The determined young girl stood in front of her mother with the younger children lined up behind, all of them clutching their cases.

'We're leaving home,' Catherine said, gravely.

'Why?' Cath asked.

'We don't think you're a suitable mother.'

And Catherine led her siblings across the yard to an old weatherboard cottage they used as an office.

'We never got to the point of working out what the unsuitable thing was.'

Cath was laughing but I couldn't help thinking that after the loss she'd endured and the strict upbringing she'd had, such blunt criticism from her own child must have been painful.

The many challenges of running a farm and raising four children as a single parent were exacerbated three years later when Benalla suffered the worst floods in living memory.

*

It was a Sunday, 3 October 1993, and heavy rain had fallen all day. The steady, continuous downpour had drenched paddocks at Yarallah and run off into the creek, which quickly broke its banks and flooded low-lying areas. The Baddaginnie Goomalibee Road was soon underwater, although that wasn't unusual – it had happened before and once the rain stopped any floodwater would usually subside quite quickly. That night the rain kept falling.

Streets in the nearby town of Benalla were soon awash and residents went to bed listening to the sound of rain drumming on their roofs. From seven o'clock on Sunday morning to nine o'clock on Monday 165 millimetres of rain fell on Benalla – about one third of the town's annual total rainfall. Residents woke to find the town submerged.

According to Cath, emergency services could have sounded the alarm but chose not to, fearing people might have panicked in the middle of the night. As a result, some people stepped out of bed and found themselves knee-deep in water.

There was no danger the hilltop house at Yarallah would flood but Cath was deeply concerned about her sheep. Some of the ewes were lambing on the home farm and she had a flock of 1200 hoggets (young sheep just beyond lambs) on an irrigation block twenty kilometres away that John had purchased before he died. The block, on the edge of the Broken River, had flooded with water rising from Cowan's floodway and spreading across the entire farm.

Cath and the children got up at dawn and drove out to the block, accompanied by Howard, the farm manager. In the pre-dawn light they could see sheep struggling to move through the rising floodwater.

Cath dropped Howard, Catherine, Charlie and Hannah on the eastern side of the farm. 'Get the hoggets to move towards the green. Lead them towards high acre,' she urged. The acre she was referring to had been fenced off since the early days of settlement and had never flooded. Cath knew that if they could just get the sheep there they'd be safe.

'Push them towards higher ground,' she shouted. 'I'll take Tom and come in from the Midland Highway.'

The farm manager plunged through the floodwater, cutting fence lines to give the exhausted animals a better chance of swimming in the same direction as the water flow, and twelve-year-old Catherine splashed through water up to her knees beside him. Farm dogs jumped along with her.

Cath and Tom approached the flood from another direction. Eight-year-old Tom had already lost his footing in a culvert and he was soaked before they even reached the main body of water.

'Mum, the lambs over here won't swim,' Catherine shouted, fighting back tears as she tried to push the terrified creatures through water that was now waist-deep. 'They won't swim!'

Cath could see the young sheep were exhausted and she knew they wouldn't survive.

'Sweetheart, they're not going to make it,' she said, as gently but as firmly as she could. 'You have to leave them. You have to concentrate on the stronger ones, we might be able to save them.'

The floodwaters were rising so fast the 'green' was disappearing in front of their eyes. Further west, the acre they were heading for turned out to be knee-deep in water as well.

'The silage mound!' Cath shouted, spotting a glimmer of hope. 'Head for the silage mound!'

Agricultural students from Dookie College drove out to see if they could help and their arrival lifted everyone's spirits. The students joined the rescue effort with renewed shouts as they drove the sheep on, pushing them towards the only clear piece of ground for miles around. Cath couldn't remember how long they worked. Time wasn't measured in hours, it was measured in lives saved – around 200 on that exhausting day.

At the end of the rescue effort their neighbour Jeannette took the children for a hot bath and Cath summoned the last of her energy to push through chest-high water with Howard and reach two sheep dogs that had survived by swimming through the water then lodging themselves in the branches of some of the trees that Cath and the children had planted.

It took four days for the floodwaters to subside and when they did the full extent of the tragedy was revealed. They lost almost 1200 sheep, 1000 of them hoggets at the river block and the remainder old ewes and new lambs from Yarallah. Clearing up the carcasses was an arduous, distressing job, shared by many.

'Australians are wonderful at pulling together in an emergency.'

Well-meaning people suggested Cath send her children to counselling to help them get over the trauma of what had happened but Cath scoffed at the idea. She had no intention of sending the children to a shrink.

'Life is full of joy, just as it is full of the potential for things out of our control to go horribly wrong,' she told her well-meaning friends.

I disagreed with Cath's dismissal of counselling yet who was I to judge? She and her children had seen firsthand how life could unexpectedly dish out the harshest blows and they had also seen how it could deliver the strongest love. The 1993 flood was just one example of how the family stood together in times of crisis.

'The young man on duty in that flood emergency couldn't make a decision and take responsibility for warning people. That showed me that I needed to bring up the children with the confidence to make tough decisions.'

Such strength of character sounded admirable, if a little fierce.

Afternoon wore into evening and the rain was still falling as we segued from tea to red wine and cheese and biscuits. Cath's combination of authority and wisdom was borne of considerable self-reflection and extensive reading (a lot of her sentences began with, 'I read something the other day . . .'). It was a beguiling combination.

'Why do we pass responsibility for living to someone else? We eat processed food bought from a supermarket that's been grown quickly, force-fed chemicals and stored for months. At least living in the country we have more capacity to grow our own.'

She was right, of course, and living in a city was no real excuse. How many of us plant lawns instead of cabbages or beetroot, decorative ferns instead of productive fruit trees?

We wandered through random topics, occasional sparks lighting the touchpaper.

'Food labelling!' she said at one point. 'Large supermarkets buy pork from China, where there's no accountability for pesticide and herbicide use, then repackage it in New Zealand so they don't have to say anything about its origin. Food comes into our country by the back door and supermarkets make money at the expense of our health. It's an abuse of the system by large corporates.'

I steered the conversation back to the topic of parenting.

'Ah yes, I was going to tell you about Penny.'

Fun was in short supply in the immediate aftermath of John's death and the demands of running a farm and raising four children left little space for it, until one of the gifts that Cath treasured – intangible and infinitely more valuable than anything she could unwrap and stick on a shelf – turned up in the shape of a neighbour's daughter. Penny offered to look after the children for a week while Cath was sent on a much-needed break, enforced by her concerned mother.

(When I heard that I revised my opinion of Kay Fairley, who up until then had assumed Cruella De Vil-like proportions in my mind. Kay lost her husband when Cath and her brother were small and she'd been forced to quit the farm to make her own way elsewhere as a single mum. Kay must have known exactly what Cath was going through.)

While Cath was away, Penny took the children into a realm Cath had never felt confident to enter – the realm of make-believe. Starting with a teddy bear's picnic, Penny encouraged the children to let their imaginations go wild. They went on a bear hunt then they were bears climbing trees. Penny spun the tale out and taught the children how to sew; she helped them make vests and jerkins out of old scraps of denim and they re-enacted the bear hunt in costume.

'And then they made a film!'

Cath came back from her break and was astonished to see a home movie of Tom, Hannah, Charlie and Catherine running through the woods, climbing trees and stalking on their hands and knees along muddy paths growling like bears.

'My mother believed that when children were growing up they should be seen and not heard.'

I struggled to hold onto my revised opinion of Kay but who was I to judge? I didn't have any children.

'Instead of keeping the children confined, as I was taught, Penny unleashed them into a world of games and fantasy.'

The young helper showed the children how to have fun and that, according to Cath, was 'an immense gift'.

The best part was that Cath could join in too. The learning unlocked something that had been missing from her own childhood and from then on she tried to include an element of fun in their daily chores and tasks.

The children often made their own entertainment, raiding the local op shop for wheels and frames so they could build and customise pushbikes; they packed picnics and took off together for the day; they dragged logs into a paddock to create a cross-country course for the ponies; they played football down the corridor that led to their bedrooms; and in summer, when it was light until nine or ten o'clock at night, they would climb out of their bedroom windows and disappear, heading for another adventure fishing, canoeing or mud-sliding at the dam.

Chores were inevitably part of their daily routine and there was pocket money to be had for bigger jobs – two dollars an hour for the easy work of checking stock (something that could be done on horseback or motorbike) and five dollars for drenching or cleaning the shed before shearing.

Tom could cook a full roast dinner by the time he was nine and Charlie was taught to slaughter lamb for the freezer by the age of twelve. Clothes were unisex so they could all be recycled and when Cath took a sewing class she made a batch of outfits all the same size; they were too big for Tom and too small for Catherine but the children wore them anyway. When Charlie outgrew Hannah, Tom inherited her cast-offs.

The children were close in age and they rarely fought. It was one for all and all for one on the Marriott family farm and big jobs like shearing always took place in the school holidays, so everyone could pitch in.

Early one winter's evening, when it was dark outside, Cath took a call from the head of a shearing team.

'We'll be there in the morning,' he said.

Cath was taken aback. She hadn't expected him for a few days and the lambs were still in the paddocks.

'I'm sorry but we haven't got the sheep in.'

'Well, you'll have to go and get them.'

The children swung into action like four musketeers.

Catherine and Hannah grabbed their ponies, Charlie and Tom jumped onto their motorbikes and together they rode over to the quarry paddock, dodging rocks to round up the sheep. By next morning, the shearing shed was full.

Hours worked were noted down and ticked off by Cath whenever she handed over payment, otherwise one of them would sidle up and suggest they hadn't been paid.

'Child labour was alive and well,' Cath joked, then she added a more serious note. 'This farm only survived because of the energy and input of everyone involved, from a very young age.'

What about Cath, I wondered, how did she survive? With no adult company she had little chance for intellectual discussion or even of committing to a community group.

'It was one of the many disadvantages of losing my husband so early,' Cath said with a total lack of self-pity. Set against that was the delight of being with her brood. Her confidence as a mother gradually increased as she marvelled at the joy of raising children.

'They were such good company. I'd been doubtful of my abilities so having children was the greatest gift. It justified my life as a person and gave me a vision and passion for the future.'

The children went to primary school in Baddaginnie, a tiny one-horse town ten kilometres away. They often rode (which must have sent the equine population soaring, sorry, couldn't resist) and a church paddock opposite was the perfect place to rest the ponies until it was time to ride home again.

Knowing there was a back road into town, Cath encouraged them to make their own way there. It was another way she taught them the value of responsibility.

'The only real danger was crossing the Melbourne to Sydney train line, especially since Tom's pony, Mitch, had a tendency to bolt.'

That sounded like a very real danger for a five-year-old but then I can't ride a horse.

'There were no lights at the crossing in those days but Catherine was a particularly good rider and so was Hannah so I knew they would look after him.'

When they weren't on ponies they rode pushbikes, and Charlie demonstrated his love for his siblings one day with a generosity of spirit that could still reduce Cath to tears.

Charlie was around eight or nine when heavy rain flooded the creek they had to cross in order to get to school. Stepping up to be the man of the family, Charlie offered to go first to see how deep it was. He got drenched.

'Well there's no point all of us getting wet,' he said.

Leaving his bike on the other side, Charlie strode back through the knee-high water and pushed first Catherine's then Hannah's and finally Tom's bike through the flood. Then he waded back and gave each of his siblings a piggyback.

Cath's voice thickened as she recounted the story and I sensed a mixture of wonder and joy at the loving generosity that lay behind such actions, coupled with sadness that young Charlie had shouldered such fierce responsibility for his siblings.

'He was such a carer,' she said. 'Charlie was my constant concern, he just didn't fit the system and he seemed so vulnerable.'

In the next breath Cath told me how much Catherine the non-conformist had worried her, especially when she took on too much responsibility, and how worried she'd been about Hannah who was a quiet achiever and easily put upon, and Tom, the youngest, who did the minimum at school to get by . . . she had clearly worried about them all.

Making sure the children were disciplined enough to 'fit in' with the rest of the world was often on Cath's mind and, with hindsight, she could recognise the fear that lay behind her own upbringing.

'Raising children without a father is not an easy thing to do.'

When Catherine cut Hannah's hair, Cath tried to believe it was to make her little sister look pretty. It was less easy to understand why Charlie would cut all the manes and tails off his sister's Little Ponies, and Cath was so angry at what she assumed was his uncaring attitude that she picked up the Coca-Cola clock radio he treasured and smashed it. Her action taught Charlie to take more care of other people's possessions yet the memory of the violence inherent in that long-ago act could still make her feel ill.

The appearance of lice horrified Cath, partly because of the classic 'I'm a bad mother' reaction and partly because the active ingredient in the shampoo the school recommended was aldrin, a toxic chemical closely related to the dieldrin they had fought so hard to avoid spraying under the house.

Cath found an old *Reader's Digest* on herbal remedies and she mixed up a brew of essential oils of rosemary, lavender and citronella. Three hours later, with everyone's head tightly wrapped in cling film, the lice were dead and a follow-up of tea-tree oil shampoo ensured they never came back.

Knowing I was planning to visit, Cath asked her children what was it about the farm that they enjoyed. Charlie's answer struck her as 'the wisest'.

'It was the space and lack of confinement on time,' he said.

She also asked them if they felt there was anything they'd missed out on being raised by a single parent (another good question and I wish I'd thought of it).

Tom's answer reassured her.

'I always had the feeling when I went to school on a Monday morning that the other kids must have done really cool stuff at the weekend that we weren't part of. Then I'd go and stay with one of them and I reckon our life was much cooler than theirs. We did way more interesting stuff.'

As Cath pointed out, life is so often a perception of what we think rather than the reality. 'We are all fumbling our way at various speeds towards the finishing line with the best intent,' she said.

It was late by the time we sat down to dinner (Moroccan stew sprinkled with seeds from a pomegranate picked from the garden that morning) and even later when we finally stopped talking. Before we packed up for the night I asked Cath if she'd ever thought of selling.

'Well I was told by a lot of people that I should.'

It was something she never contemplated, in spite of mounting bills she often couldn't pay. When cash was short she would ring local businesses and explain there wouldn't be any money for at least three months. Their answer was always the same.

'That's fine, Cath, just keep doing business with us.'

And she did.

'I've always shopped locally and I always will. That's what helps ensure jobs for other children in town in the future.'

With high school looming, Cath had to make a decision. She could send the children to a local school and keep working to pay off the farm debt, or she could invest in their education. Education won.

Even with the generous discount Scots, in Albury, offered Cath for having four full-time boarders, the fees were considerable. In the end she was forced to sell the irrigation block where they had planted 10,000 trees in anticipation of the day when the children might need building materials of their own.

'I don't regret it,' she said firmly. 'I chose to invest in school fees because education and respect for others is all you can really hope to give children in life.'

It seemed a fitting place to end for the night.

The rain had cleared by the following morning to reveal a

landscape of startling beauty. I stood on the veranda looking across swathes of jonquils nodding cheerful heads above a carpet of statice. Primrose and cyclamen flowered in profusion and the view across native grassland and green paddocks dotted with sheep slid gently down to a creek bed, its outline marked by a line of newly planted shrubs.

There was a tangible sense of energy, as if the earth itself exuded power, and I took a deep breath. There's surely no better time to be outdoors than just after it's rained.

'Sending the children to school meant they could participate in sport and music,' said Cath, handing me a bowl of porridge made with natural yogurt, honey and activated almonds.

'I couldn't have ferried them to four different sports activities and be running the farm at the same time,' she said, adding a pot of tea and a milk jug to the round table, which was sheltered from the brisk wind.

'How could you not have regard and respect for this?' she asked, taking in the sweeping views. 'Not as a chocolate box but as part of your essence?'

Absorbing the sights, sounds and smells of that freshly washed land it was easy to appreciate what Cath meant.

A king parrot preened itself in a nearby tree as we picked up the conversation from the night before.

All four children went to Scots, an hour away from home, and Cath's instinct proved spot-on. Catherine, Hannah, Charlie and Tom all played for the school in various sports and they all got involved in music too, albeit with varying degrees of success.

Hannah composed a seven-part drum fanfare that won the eisteddfod in Albury one year and she was drum sergeant in the school pipe band; Catherine was a front-row piper; Charlie was a gifted musician who took up the cello and then went on to the chanter (even though his real interest was working at home on the farm); and Tom took up the French horn at primary school, giving it up soon afterwards.

'Like his darling father, Tom was tone deaf and couldn't sing a note in tune.'

If Cath had worried about the children when they were at primary school she worried about them even more in their senior years. It was all such an unknown. How would they turn out? How would they manage their lives? Were they in the right environment to maximise their talents?

By the time I met Cath her oldest daughter Catherine was a successful businesswoman and passionate advocate for women in rural areas. At school Catherine had been less focused.

'She caught up with her year eleven senior recently and Yvette said, "Do you remember me?" At school Catherine had been so embarrassed. She said, "I'm sorry I led you the most awful dance." Yvette said, "Yes, you did, and I'm glad to see how well you've done. I actually thought you'd end up in prison."'

Cath gave a big laugh and was quick to point out that none of her children took drugs or committed any crime.

'It was just that they weren't going to toe the line in a system if they didn't believe in it.'

The system didn't nurture Charlie's particular talents so Cath took him out of Scots in year eight and sent him to St Paul's College, a Lutheran farm school in Walla Walla. The school encouraged Charlie's interest in agriculture and helped him discover a remarkable gift for fixing broken machinery and hand-ling animals at local shows. As a result, Charlie dumbfounded himself (and his mother) by sailing through his school certificate.

Catherine, Hannah and Tom all chose to study for a bachelor of rural science at the University of New England in Armidale and Charlie opted for a diploma of agriculture at Longreach Pastoral College. Letting them go was much harder than it had been when they'd gone to school. Armidale was an eleven-hour drive and Longreach was twenty-two hours away. The children were growing up and getting further away from Cath in every sense.

'I had a down period when they left. They'd been such good company.'

Cath made the eleven-hour drive to Armidale several times a year and twice she drove up to Longreach to visit Charlie. At one of the formal prize-giving dinners the name Charlie was on everyone's lips.

'Who's Charlie?'

'Have you seen what Charlie did?'

A guest approached Cath and asked if she was Charlie's mother. She nodded warily. 'Is there something I should know?'

The guest smiled. 'I flew in from Brisbane today. When the plane came in to land I saw the name "Charlie" in massive letters out on a claypan.'

Cath wasn't surprised to learn that Charlie had carved his name on the claypan while doing his grader driver's ticket. It wasn't the first time it had happened; as a young teenager he'd used his motorbike to carve his name out of the grass in a paddock at Yarallah.

'Why did you do it?' Cath had asked him when she saw the churned-up grass.

Charlie shrugged. 'It gives me something to think about while I'm away at school, knowing my name's part of the farm.'

Cath's oldest son had always said that he wanted to take over the farm so Cath didn't feel too bad when she dragged him away from a young man's dream world of roo-shooting in Queensland with a mate who drove a V8 ute.

'Kangaroo shooting and a fast car seemed a dangerous mix for a nineteen-year-old.'

Cath sensed that if she stayed around dispensing motherly advice she would suffocate her son, so in 2002 she took a twelve-month job in community development in nearby Shepparton, followed by another year running the Goulburn Murray Community Leadership Program.

She worked away from the farm for two years in all, knowing that Charlie needed the chance to prove himself. He would have done too, if it hadn't been for the crippling drought that scuppered the innovative plans he'd put in place.

Drought hit the Goulburn Valley hard in 2003 and water allocation on irrigated land that season was set at six per cent instead of the normal hundred per cent. In 2004, they went from an average annual rainfall of twenty-five inches down to just seven, and with no grass for their 4000 sheep Charlie was forced to buy in expensive fodder then off-load stock cheaply. With no male role model to help, and a drought that refused to break, the difficulties he faced were immense.

'You can't finish stock or breed on land when it doesn't rain and there's no grass. The mistake I made with Charlie was leaving him to manage the farm when it was too big and he was too young.'

It was typical of Cath that she saw it as her failure and she was quick to acknowledge Charlie's contribution in setting up a successful lamb finishing operation that continued to this day. 'And he put in place systems that could be modified to meet the changing climate.'

Many farmers were devastated by the drought and Cath realised through her work in community development that some people had real trouble with change.

'They wanted to continue with the status quo, regardless of the fact that there was no water in storage and no rain. It didn't seem to register with these farmers that there would be no water available that season.' The experience prompted her to apply for a course at Swinburne University, a masters in strategic foresight.

Hannah had done an honours degree in rural science, with a special focus on feed intake and growth, and she chose to work on Yarallah to better understand the lamb finishing business. Having worked on properties in western Victoria and South Australia, where intensive feeding systems were being run profitably, Hannah's focus and attention to detail helped build their reputation for lamb that was always finished to exact specifications and could command a premium price.

When Charlie moved to Brewarrina to work it gave Hannah the opportunity she craved to gain more experience in her chosen field. Cath's second daughter was extremely well organised and she stayed at Yarallah for four years, concentrating on systems and sharing some of the work of running a big business alongside Tom, who concentrated more on stock. Hannah had since moved on to manage a large property north of Melbourne.

Tom was managing Yarallah now and his focus was on developing a strong biological content in the soil. He had developed a system that would continue to be sustainable, even with the prospect of a changeable climate, and when I visited he was about to embark on a course in alternative energy systems through the University of Western Australia.

Cath and Tom were both passionately committed to maintaining the ground cover on Yarallah and working to establish a system of farming that would cope with whatever rainfall they received, whether it fell in winter, autumn or spring. Their ideal mix was a blend of ryegrass, clover and native perennials. Getting the balance right was what kept them farming.

Cath's children had all inherited a desire to keep learning and to keep improving. 'Each of them is passionate about farming,' she said.

Passion nourished Cath's soul.

'Life can get tough. If you have nothing to fall back on as your strength or your saving grace you can get into a black situation very quickly. You need something to be passionate about.'

And what of Catherine, her oldest daughter? Catherine left home to work as a jillaroo in Townsville before qualifying as a ruminant nutritionist and taking work in Malaysia and Indonesia then Brisbane, Perth, Broome and Canberra – it was hard to keep up. She was a Myers Briggs practitioner, a part-time wedding photographer and in 2012 she was named West Australia Rural Woman of the Year. That same year she founded the organisation Influential Women to address the gulf of misunderstanding between people who grow food and people who consume it.

'Catherine says people don't want to know what you know, they want to know how much you care.' (Transcribing the interview tapes I underlined that sentence twice.)

'People don't care if you want to be a farmer or not, they care if you're doing a good job and if you're caring for your animals and the land and environment.'

Well said and Cath was a shining example of that.

'At the end of the day it doesn't matter if you have ten cows or 10,000,' Cath added. 'It's about the contribution you can make to the food chain.'

Cath agreed with her daughter about the need to improve communication between the farming and non-farming sections of society.

'It's not a question of telling people how much it hurts if you buy milk at a dollar a bottle [it does by the way, in case you're wondering], it's just that farmers can supply healthy food so if you're concerned about what you feed your family you shouldn't always pick the cheapest option.'

Cath fixed me with a look that got my attention.

'And you have to use language people understand, so for example there's no point me talking to you about mulesing a wether.'

I sighed. After eight interviews my near-total ignorance of farming was still impossible to hide.

In the twenty-nine years that Cath and her family had farmed their 900 hectares near Benalla, their aim had always been to improve the land. Cath planted trees on a hillside that was once a quarry, cladding the bare clay in sixteen hectares of grassy woodland, and up to ten per cent of the farm was now covered in trees and perennial grasses with woody, understory species lining the creek.

It seemed like a random act of generosity that added nothing to the bottom line but in fact there was a clear rationale behind the improvements.

'The birds get more habitat and they eat the bugs that eat the grass so we don't have to use any pesticides.'

Rotational grazing gave paddocks time to recover and a weak solution of round-up, mixed with sugar solution to feed microbes in the soil, was all Tom would ever use to knock back poor nutritional grasses.

'I would never say that using chemicals is wrong, I know it's the only way some farmers survive,' said Cath. 'We choose not to do it, that's all.'

The weather was so mild that afternoon we took a long walk through the garden, starting in an unexpected rainforest that somehow flourished in their Mediterranean climate on top of a shallow clay hill in northern Victoria.

There were towering red cedars from Coffs Harbour, leopardwood, *Flindersia Australis*, African lilies, *thunbergia gardenia*, tea plants, cardamom, orchids and many more exotic and unusual species, all nurtured under a thick layer of sheep's wool, straw and wood mulch.

Pear trees planted by Cath's mother produced kilos of fruit each year, as did the pomegranate, guava, persimmon, nectarine, fig, orange, lemon, grapefruit and apple trees Cath had planted. Cockatoos normally got the almonds before Cath and they were partial to the quince as well.

'White cockatoos are what I dislike most about life in general,' Cath muttered then acknowledged in the next breath that white cockatoos were nature's pruners, creating the leaf litter that enriched the soil. Cath could always be relied on to see both sides of an argument and then to focus on the positive.

Someone once suggested to Cath that she stop worrying about the forest and try saving a single tree.

'But I can't conceive of a tree that's not part of a forest. That's my limitation.'

Any visitor who (like me) delighted in her extraordinary garden would surely consider it another one of her precious gifts rather than a limitation.

Cath never lost the love of books she developed as a child and she was still an avid reader. As we made our way through a dry sclerophyll forest and across the grassland beyond, Cath recalled reading an excerpt from the journal of James Augustus Robinson in 1839. It spoke of vast plains in northeast Victoria with waist-high grass, green at its base, and waterholes teeming with fish.

'This was in February, when temperatures were 110 Fahrenheit every day. Victoria once had some of the most diverse, deep-rooted grasslands in the world, with anything from eighteen to twenty per cent carbon in the soil. We'd be lucky to have one per cent in most Australian soils now.'

Not at Yarallah. The ground beneath our feet was springy and moist. It looked and felt like nature at its most powerful.

'I like people to come and just be in nature,' said Cath, clearly pleased that I was enjoying her garden.

We ended our walk on a track that circled the sparkling waters of a large dam – their only source of water save what fell from the sky – which Cath revealed was partly a sixtieth birthday present.

Charlie borrowed a bobcat then a digger and used the powerful machines to heap up clay near the dam, then he and his siblings created a path along the mound of clay using mulch from pine trees they'd felled. The path led to a place where Cath could sit and contemplate the beauty of nature, looking across the dam that never emptied, not in the worst drought. They'd even made steps down to the water, so she could go swimming.

There was never any money for gifts when the children were growing up so they all became adept at making presents. Hannah built her mum a hidden fairy den one year, painting red and yellow spots on upturned wicker baskets to create mushroom stools which she placed in a circle under the casuarina trees near the dam. Cath was led down to the fairy den at nine o'clock that night to celebrate.

And there were surprise appearances at each other's birthdays, no matter how far they had to travel.

On her twenty-first birthday, Catherine was in Moree in northern New South Wales, bug-checking cotton in the university summer holidays, so Cath, Tom, Hannah and Charlie drove up to join her.

Carrying swags, a bag of provisions, a slab of beer and a box of utensils they traipsed into the bush, searching for a suitable site to make camp. The first spot was taken so they picked up their provisions and kept going, clambering over rocks and using ropes to drop down into a clearing just as it got dark. That's when they realised they'd forgotten to pick up the utensils.

Cath woke the next morning to the smell of a cooked breakfast and she opened her eyes to see Charlie had buried a flat rock in the glowing embers. He was frying bacon and cooking toast on the rock and Tom had used his Leatherman to turn up the edges of a beer can for fried eggs.

'It was the most delicious breakfast any of us had ever tasted.'

Hannah also had to work on her birthday so she assumed nothing would happen. She should have known better. Cath and Tom drove to the Melbourne markets at six in the morning to buy fruit and turned up at her farm for breakfast, joined shortly afterwards by Catherine who'd flown in from Indonesia and Charlie and his girlfriend Pip, who'd been working out of Broome.

Catherine held her thirtieth in January 2011 at Marble Bar, arguably one of the hottest places in northern West Australia, and she was told only Tom could make it. Then Charlie stepped off the same plane and Hannah turned up at the fancy-dress party later that night disguised as a Hills Hoist.

They may have gone their separate ways at university and beyond but strong ties still bound them together.

Cath was a great believer in energy, the kind that was available from nature and the universe when she needed it (which wasn't necessarily when she wanted it), and standing in that magical place near the dam I had no problem believing Cath when she said that John's spirit was still around. The knowledge that he was somehow connected to her and the family helped motivate Cath.

'John and I dreamt that we would have the farm for the children when they decided what they wanted to do with their lives.'

She smiled.

'Well, we still have the farm and it has done a lot to contribute to the lives not only of our own children, but others that we have known too. We've all gained from it and learned so much about caring, about initiative, resourcefulness and responsibility, and it wouldn't have happened any other way that I know about. Equally I know I've got things to finish yet.'

One of those is her new house.

*

It was late afternoon on Yarallah, a hot summer's day in February 2011, and Cath and Hannah were moving a mob of ewe hoggets back to a difficult gateway. Hannah looked up from their work for a moment and spotted a drift of smoke that seemed to be coming from the direction of their house.

'Mum, can you see that?' she said.

Cath turned to look at where her daughter was pointing. The wisp of smoke was clearly visible against the cloudless sky, and in a matter of moments it thickened, darkening ominously. Cath and Hannah both knew there were no controlled burns in the area.

'We'd better get back and see what's happening.'

Hannah grabbed the young sheep dog she'd been training, Cath swung the bike around and they sped back up the track. Halfway up the hill Cath reached for her mobile; the fire was definitely coming from their house. She dialled triple zero. By the time they pulled up outside the house black smoke was billowing through the main room. Cath opened the side door and the flames exploded in a rush of oxygen.

Hannah raced to the kennel to secure the pup then she ran for the farm trailer with the Goomalibee fire unit. At the same time she phoned Tom, who was down at the other end of the farm.

'There's a fire at home, you'd better get back'.

Tom had assumed it was a small fire so he was shocked when he arrived to see the whole roof ablaze. In an extraordinary coincidence the head of the local fire brigade, Captain Peter Bailey, arrived at the same time. He'd been on his way to Yarallah to let Cath, Tom and Hannah know that that night's meeting of the Goomalibee Country Fire Authority had been cancelled and he received the triple zero call as he drove up their tree-lined approach road. He pulled up outside in time to see the windows explode.

In all, six fire crews fought the blaze that night, from Benalla, Baddaginnie, Goomalibee, Koonda and Upotipotpon. When they ran out of water they filled up from the dam, hosing the house from every direction in an effort to douse the flames. At the end of the night all that was left was the bedroom wing, badly smoke-damaged but still standing. The fact that the bedroom wing still had a roof the next morning was entirely due to the Goomalibee fire team of Peter Bailey and 'Barney' Button, who camped on the lawn to keep watch. Their vigilance was rewarded when embers re-ignited in the middle of the night and they were able to extinguish them.

Morning revealed the full extent of the damage. The home Cath and John had built was an ash-filled relic. Cath lost nearly everything in the fire, including tables, chairs, shelves and salad bowls she'd crafted herself over decades at weekly wood-turning classes. A bone china coffee set Cath had inherited from her grandmother turned to ash in the intense heat and all the pictures on the walls were destroyed.

Her treasured collection of books went too, some of them irreplaceable volumes that she had collected from her time at the BBC, others dating back to the 1730s. There were early Australian volumes on Aborigines, plants and history, as well as her mother's collection of books on plants. Many of them were short print runs or books from her travels that could never be replaced.

Most distressing of all was the loss of a framed collage of family photographs that had been presented to John on his fortieth birthday, a few months before he died.

I suggested to Cath that some people seem to suffer more than their fair share of tragedy (although what's 'fair' about tragedy?) and she shook her head.

'We've had major challenges but then you think about people who went to war. Some people survived and came home and some of them didn't. It's about getting a perspective on life.'

Cath, Tom and Hannah picked through the waterlogged remains of blackened timber and ash, finding random pieces that oddly didn't burn, like a silver teapot from Andrew, Cath's brother. Some of the photographs didn't burn either. Those trapped in the middle of bulging albums and squashed under the weight of heavy books acted like dense logs and burnt only on the perimeter.

Cath moved to a rundown cottage nearby that stood within sight of the burnt-out shell. She found the view from the window difficult to cope with.

Friends who heard about the fire turned up with random gifts. Cath recalled one friend, Hilary, arriving with several pairs of silk curtains and a coffee machine. Her much loved younger brother, Andrew, found paintings at Aingers auction rooms in Richmond and he gave her furniture too. The children threw a dinner party and asked everyone who came to bring a book.

People said it would take six months before Cath could move back into her ruined house; in the end it took two and a half years and she was still working on it when I visited.

Trees and plants growing next to the house that were scorched in the fire were uprooted and replanted and Cath was relieved that she didn't lose too many.

'Nature has this wonderful ability to survive.'

The rose garden planted in John's memory was saved and the roses bloomed as a new house gradually grew out of the ashes. It wasn't that different from the old one, apart from a higher roof line to change the angle of clerestory windows, a wide covered deck and more doors in the corridor that led to the bedrooms. If the fire had started at night, Cath realised later, she would have been trapped.

The long kitchen bench in her new house was topped with timber cut from the property and the walls were painted a textured ochre to conjure a sense of enclosure – like a cross between a Spanish adobe and a North African mud hut. Cath brought the outside in with a full-length window in the bathroom that over-looked her rainforest garden.

'No one walks around that side of the house so they're unlikely to see in.'

Cath probably wouldn't care too much if they did.

One of the gifts Cath treasured was her new bed, a whimsical fancy of steel, corrugated iron and pieces of old farm machinery that her children made for her. Another was a painting that hung on a wall in the living room, given to her by her thoughtful brother to make up for the many she lost. The abstract image of a bushfire racing up a forested hillside had a small patch of blue sky just visible in the top corner. When I looked closer I noticed the wind had veered, suggesting the small stand of trees at the top would be spared.

The picture's innate sense of optimism seemed to echo Cath's attitude to life.

Cath treated her animals with the respect she accorded every living thing. Knowing they would end up on someone's table, she nurtured them, acknowledged their contribution to the food chain, and made their end as swift and as painless as possible. Her sensibilities were New Age but she wasn't about to turn into a hippy fruitarian anytime soon.

'It's about connection. It's about respecting the littlest things that are all part of this great system we belong to.'

We talked about succession and she laughed.

'When I die it will be fine. The problem will be if I live a long time and hang around.'

Notwithstanding the years of effort Cath had put into creating a place of natural beauty she had no particular attachment to that piece of land, or even to that part of Australia. She could happily move on, to another piece of Australia where she would produce food for the future.

'I could just as easily sell up if the kids aren't interested in staying here. What matters is creation, not ownership.'

Cath was no fool when it came to finance; she knew the farm had to turn a profit. She also knew money didn't buy the real rewards in life.

'It is the learning that brings understanding, that is the greatest reward.'

Tom had learnt to approach his work on Yarallah in a similar way. He knew he wouldn't be there forever, and he knew he only had a quarter share in a farm that would probably be sold one day, which didn't stop him mending fences and getting stuck into renovating the weatherboard cottage that offered holiday accommodation on Yarallah.

'It doesn't matter if the farm is mine or not, it will still be better if we improve on what we started with, and the knowledge that I gain is beyond price,' Tom said.

Cath used to dream that her children might use Yarallah as a base, each of them exploring their own particular talents to create a joint venture. 'We've got the housing and the infrastructure to run any kind of business,' she mused. 'They could rent out the cottages, grow vegetables, run a grazing enterprise, expand the lamb finishing business . . .'

She stopped herself and shook her head. The dream was hers, not theirs, and she had long since accepted that her children's lives had taken a different path, even if only for the moment. Whatever they did, and wherever they lived, their connection would remain strong.

She quoted from Khalil Gibran's *The Prophet*.

'Your children are not your children. They are the sons and daughters of Life's longing for itself . . .'

And there was that life-giving rain again, falling from a clear blue sky.

References

Books and articles

Andrews, Alison, 'A rare woman in a man's world', *The Examiner*, 24 October 2013

Barrier Miner, *Died while playing cricket: Part Owner of Kara Station*, Broken Hill, 11 June 1934, www.trove.nla.gov.au

Bowen, Jill, *Kidman: the Forgotten King*, Fourth Estate, 1987

Brown, Felicity, 'Mover and Shaker', *RM Williams Outback magazine*, Issue 84, Aug/Sep 2012

Brown, RB, *The desertion of Gilberton*, in Dalton, BJ (Ed) Lectures on North Queensland history, Townsville, James Cook University, 1974. www.espace.library.uq.edu.au

Department of Environment and Primary Industries, *Breeds of beef cattle*, www.depi.vic.gov.au

DPI, *Roma Britnell, Australian Rural Woman of the Year*, 2009, Rural Women's Awards, Department of Primary Industries, www.dpi.vic.gov.au

Freer, Monique, 'Benalla flood remembered', *Benalla Ensign*, 4 Oct 2013, www.mmg.com.au

Gabbrielli, Felicity, *BP Ekberg: Sweden to Australasia*, Cremorne1. com, New South Wales, 2011

References

Gibran, Khalil, *The Prophet*, Pan Books, London, 1991

Martin, Jess, 'Kidman's dreams live on', *RM Williams Outback magazine*, Issue 80, Dec/Jan 2012

Mountford, CP and Ainslie, R, *The Dreamtime*, Griffin Press, Adelaide, 1965

Nally, Steve, 'Dairying to be different', *RM Williams Outback magazine*, Issue 75, Feb/Mar 2011

Payne, Alan, 'The Legend of Arnold Green', *Naval Historical Review*, March 1977, www.navyhistory.org.au

RIRDC, *Taking kids safely to work on the farm this Christmas*, Rural Industries Research and Development Corporation, 8 Dec 2014, www.rirdc.gov.au

Rivet, Rohan, 'Victorian farmers flayed by drought and falling prices', *The Canberra Times*, 12 May 1976, www.trove.nla.gov.au

Sturmer, Jake, *Union raises fears that Telstra communication pits contain deadly banned pesticide dieldrin*, ABC News, 6 June 2013, www.abc.net.au

Sullivan, Kath, *Comb dispute three decades on*, ABC Rural, 23 July 2013, www.abc.net.au

Sunshine Coast Daily, 'Pressure can even get to the very best', 4 June 2013, www.sunshinecoastdaily.com.au

Tauman, Merab Harris, *O'Connor, Charles Yelverton (1843–1902)* Australian Dictionary of Biography, Vol 11 (MUP) 1988, www.adb.anu.edu.au

Terrell-Payne, Jane and Payne, Les, with additional material from Ken Granger and Col Grant, *.28°S 144°E Thargomindah – Queensland by Degrees*, 2008, www.rgsq.org.au

Trading Economics, 'Australia Average Weekly Wages, 1969–2015', www.tradingeconomics.com (undated, website visited 13 Feb 2014)

Wells, Kathryn, *Australian Farming and Agriculture*, updated 25 June 2013, www.australia.gov.au

References

YDCM, *Funding legislation could close day care centres*, York and
District Community Matters, June 2010, www.ydcm.com.au

Websites

www.accbeef.net.au
www.aghealth.org.au
www.bom.gov.au
www.dpi.vic.gov.au
www.environment.gov.au
www.farmz.com.au
www.gilbertonoutbackretreat.com
www.kidman.com.au
www.nationalparks.nsw.gov.au
www.plymouthbrethrenchristianchurch.org
www.rdawheatbelt.com.au
www.rgsq.org.au
www.rqm.com.au
www.visitnsw.com
www.wetlandinfo.ehp.qld.gov.au
www.yarallah.com.au
www.yarddogsnsw.com.au

Acknowledgements

I'm enormously grateful to publisher Ingrid Ohlsson for proposing this book and then for suggesting I might like to write it. Ingrid's belief, support and advice were invaluable. Sincere thanks to talented editor Sam Sainsbury, to the team of professionals at Macmillan and to Deonie Fiford, who did an excellent job of editing the text. Thanks also to my supportive agent Pippa Masson at Curtis Brown.

Many people helped me find then contact potential interviewees, especially Ruth Redfern, Kate Philipson, Mary Frost, Jeanette Gatenby, Jo Vidorin, Lynne Gall, Chris Larsen, Annabelle Brayley, Eleanor Falkiner, Allison Priest, Rowena Martin and the team at the RIRDC Rural Women's Award. I also want to thank Rob Gillam for explaining the subtle differences between a pastoralist and a grazier.

My sincere thanks to Chris Belshaw and the teams at the Royal Flying Doctor Service SE Section and Santos for generously allowing me to fly from Broken Hill to Moomba and Ballera then back to Adelaide so I could interview Michelle Reay and Jon Cobb at Durham Downs.

Acknowledgements

As always, my partner Clyde has been a constant support throughout the research and writing of this book. I would not have finished it without his help.

And last but certainly by no means least, my sincere thanks to all the families I interviewed, who welcomed me into their lives with such humbling generosity and who shared their stories so openly and honestly.